Glencoe McGraw-Hill

MW00511722

Study Guide and Intervention Workbook

Algebra 1

McGraw Hill Education

To the Student

This *Study Guide and Intervention Workbook* gives you additional examples and problems for the concept exercises in each lesson. The exercises are designed to aid your study of mathematics by reinforcing important mathematical skills needed to succeed in the everyday world. The materials are organized by chapter and lesson, with two Study Guide and Intervention worksheets for every lesson in *Glencoe Algebra 1*.

Always keep your workbook handy. Along with your textbook, daily homework, and class notes, the completed *Study Guide and Intervention Workbook* can help you in reviewing for quizzes and tests.

To the Teacher

These worksheets are the same ones found in the Chapter Resource Masters for *Glencoe Algebra 1*. The answers to these worksheets are available at the end of each Chapter Resource Masters booklet as well as in your Teacher Wraparound Edition interleaf pages.

connected.mcgraw-hill.com

 **Education**

Send all inquiries to:
McGraw-Hill Education
8787 Orion Place
Columbus, OH 43240

ISBN: 978-0-07-660292-6
MHID: 0-07-660292-3

Printed in the United States of America.

12 13 14 15 16 17 18 19 RHR 19 18 17 16

The *McGraw-Hill* Companies

Contents

Lesson/Title	Page	Lesson/Title	Page

1-1 Study Guide and Intervention
Variables and Expressions

Write Verbal Expressions An **algebraic expression** consists of one or more numbers and variables along with one or more arithmetic operations. In algebra, **variables** are symbols used to represent unspecified numbers or values. Any letter may be used as a variable.

Example Write a verbal expression for each algebraic expression.

a. $6n^2$

the product of 6 and n squared

b. $n^3 - 12m$

the difference of n cubed and twelve times m

Exercises

Write a verbal expression for each algebraic expression.

1. $w - 1$

2. $\frac{1}{3}a^3$

3. $81 + 2x$

4. $12d$

5. 8^4

6. 6^2

7. $2n^2 + 4$

8. $a^3 \cdot b^3$

9. $2x^3 - 3$

10. $\frac{6k^3}{5}$

11. $\frac{1}{4}b^2$

12. $7n^5$

13. $3x + 4$

14. $\frac{2}{3}k^5$

15. $3b^2 + 2a^3$

16. $4(n^2 + 1)$

1-1 Study Guide and Intervention (continued)

Variables and Expressions

Write Algrebraic Expressions Translating verbal expressions into algebraic expressions is an important algebraic skill.

> **Example** Write an algebraic expression for each verbal expression.
>
> **a. four more than a number n**
>
> The words *more than* imply addition.
> four more than a number n
> $4 + n$
> The algebraic expression is $4 + n$.
>
> **b. the difference of a number squared and 8**
>
> The expression *difference of* implies subtraction.
> the difference of a number squared and 8
> $n^2 - 8$
> The algebraic expression is $n^2 - 8$.

Exercises

Write an algebraic expression for each verbal expression.

1. a number decreased by 8

2. a number divided by 8

3. a number squared

4. four times a number

5. a number divided by 6

6. a number multiplied by 37

7. the sum of 9 and a number

8. 3 less than 5 times a number

9. twice the sum of 15 and a number

10. one-half the square of b

11. 7 more than the product of 6 and a number

12. 30 increased by 3 times the square of a number

1-2 Study Guide and Intervention

Order of Operations

Evaluate Numerical Expressions Numerical expressions often contain more than one operation. To evaluate them, use the rules for order of operations shown below.

Order of Operations	**Step 1** Evaluate expressions inside grouping symbols.
	Step 2 Evaluate all powers.
	Step 3 Do all multiplication and/or division from left to right.
	Step 4 Do all addition and/or subtraction from left to right.

Example 1 Evaluate each expression.

a. 3^4

$3^4 = 3 \cdot 3 \cdot 3 \cdot 3$ Use 3 as a factor 4 times.
$\quad = 81$ Multiply.

b. 6^3

$6^3 = 6 \cdot 6 \cdot 6$ Use 6 as a factor 3 times.
$\quad = 216$ Multiply.

Example 2 Evaluate each expression.

a. $3[2 + (12 \div 3)^2]$

$3[2 + (12 \div 3)^2] = 3(2 + 4^2)$ Divide 12 by 3.
$\qquad\qquad\qquad = 3(2 + 16)$ Find 4 squared.
$\qquad\qquad\qquad = 3(18)$ Add 2 and 16.
$\qquad\qquad\qquad = 54$ Multiply 3 and 18.

b. $\dfrac{3 + 2^3}{4^2 \cdot 3}$

$\dfrac{3 + 2^3}{4^2 \cdot 3} = \dfrac{3 + 8}{4^2 \cdot 3}$ Evaluate power in numerator.

$\qquad = \dfrac{11}{4^2 \cdot 3}$ Add 3 and 8 in the numerator.

$\qquad = \dfrac{11}{16 \cdot 3}$ Evaluate power in denominator.

$\qquad = \dfrac{11}{48}$ Multiply.

Exercises

Evaluate each expression.

1. 5^2

2. 3^3

3. 10^4

4. 12^2

5. 8^3

6. 2^8

7. $(8 - 4) \cdot 2$

8. $(12 + 4) \cdot 6$

9. $10 + 8 \cdot 1$

10. $15 - 12 \div 4$

11. $12(20 - 17) - 3 \cdot 6$

12. $24 \div 3 \cdot 2 - 3^2$

13. $3^2 \div 3 + 2^2 \cdot 7 - 20 \div 5$

14. $\dfrac{4 + 3^2}{12 + 1}$

15. $250 \div [5(3 \cdot 7 + 4)]$

16. $\dfrac{2 \cdot 4^2 - 8 \div 2}{(5 + 2) \cdot 2}$

17. $\dfrac{4(5^2) - 4 \cdot 3}{4(4 \cdot 5 + 2)}$

18. $\dfrac{5^2 - 3}{20(3) + 2(3)}$

1-2 Study Guide and Intervention *(continued)*

Order of Operations

Evaluate Algebraic Expressions Algebraic expressions may contain more than one operation. Algebraic expressions can be evaluated if the values of the variables are known. First, replace the variables with their values. Then use the order of operations to calculate the value of the resulting numerical expression.

Example **Evaluate $x^3 + 5(y - 3)$ if $x = 2$ and $y = 12$.**

$$
\begin{aligned}
x^3 + 5(y - 3) &= 2^3 + 5(12 - 3) && \text{Replace } x \text{ with 2 and } y \text{ with 12.} \\
&= 8 + 5(12 - 3) && \text{Evaluate } 2^3. \\
&= 8 + 5(9) && \text{Subtract 3 from 12.} \\
&= 8 + 45 && \text{Multiply 5 and 9.} \\
&= 53 && \text{Add 8 and 45.}
\end{aligned}
$$

The solution is 53.

Exercises

Evaluate each expression if $x = 2$, $y = 3$, $z = 4$, $a = \dfrac{4}{5}$, and $b = \dfrac{3}{5}$.

1. $x + 7$ **2.** $3x - 5$ **3.** $x + y^2$

4. $x^3 + y + z^2$ **5.** $6a + 8b$ **6.** $23 - (a + b)$

7. $\dfrac{y^2}{x^2}$ **8.** $2xyz + 5$ **9.** $x(2y + 3z)$

10. $(10x)^2 + 100a$ **11.** $\dfrac{3xy - 4}{7x}$ **12.** $a^2 + 2b$

13. $\dfrac{z^2 - y^2}{x^2}$ **14.** $6xz + 5xy$ **15.** $\dfrac{(z - y)^2}{x}$

16. $\dfrac{25ab + y}{xz}$ **17.** $\dfrac{5a^2b}{y}$ **18.** $(z \div x)^2 + ax$

19. $\left(\dfrac{x}{z}\right)^2 + \left(\dfrac{y}{z}\right)^2$ **20.** $\dfrac{x + z}{y + 2z}$ **21.** $\left(\dfrac{z \div x}{y}\right) + \left(\dfrac{y \div x}{z}\right)$

1-3 Study Guide and Intervention

Properties of Numbers

Identity and Equality Properties The identity and equality properties in the chart below can help you solve algebraic equations and evaluate mathematical expressions.

Additive Identity	For any number a, $a + 0 = a$.
Additive Inverse	For any number a, $a + (-a) = 0$.
Multiplicative Identity	For any number a, $a \cdot 1 = a$.
Multiplicative Property of 0	For any number a, $a \cdot 0 = 0$.
Multiplicative Inverse Property	For every number $\frac{a}{b}$, where a, $b \neq 0$, there is exactly one number $\frac{b}{a}$ such that $\frac{a}{b} \cdot \frac{b}{a} = 1$.
Reflexive Property	For any number a, $a = a$.
Symmetric Property	For any numbers a and b, if $a = b$, then $b = a$.
Transitive Property	For any numbers a, b, and c, if $a = b$ and $b = c$, then $a = c$.
Substitution Property	If $a = b$, then a may be replaced by b in any expression.

Example Evaluate $24 \cdot 1 - 8 + 5(9 \div 3 - 3)$. Name the property used in each step.

$$
\begin{aligned}
24 \cdot 1 - 8 + 5(9 \div 3 - 3) &= 24 \cdot 1 - 8 + 5(3 - 3) && \text{Substitution; } 9 \div 3 = 3 \\
&= 24 \cdot 1 - 8 + 5(0) && \text{Substitution; } 3 - 3 = 0 \\
&= 24 - 8 + 5(0) && \text{Multiplicative Identity; } 24 \cdot 1 = 24 \\
&= 24 - 8 + 0 && \text{Multiplicative Property of Zero; } 5(0) = 0 \\
&= 16 + 0 && \text{Substitution; } 24 - 8 = 16 \\
&= 16 && \text{Additive Identity; } 16 + 0 = 16
\end{aligned}
$$

Exercises

Evaluate each expression. Name the property used in each step.

1. $2\left[\dfrac{1}{4} + \left(\dfrac{1}{2}\right)^2\right]$

2. $15 \cdot 1 - 9 + 2(15 \div 3 - 5)$

3. $2(3 \cdot 5 \cdot 1 - 14) - 4 \cdot \dfrac{1}{4}$

4. $18 \cdot 1 - 3 \cdot 2 + 2(6 \div 3 - 2)$

1-3   Study Guide and Intervention (continued)

Properties of Numbers

Commutative and Associative Properties The Commutative and Associative Properties can be used to simplify expressions. The Commutative Properties state that the order in which you add or multiply numbers does not change their sum or product. The Associative Properties state that the way you group three or more numbers when adding or multiplying does not change their sum or product.

Commutative Properties	For any numbers a and b, $a + b = b + a$ and $a \cdot b = b \cdot a$.
Associative Properties	For any numbers a, b, and c, $(a + b) + c = a + (b + c)$ and $(ab)c = a(bc)$.

Example 1   Evaluate $6 \cdot 2 \cdot 3 \cdot 5$ using properties of numbers. Name the property used in each step.

$$
\begin{aligned}
6 \cdot 2 \cdot 3 \cdot 5 &= 6 \cdot 3 \cdot 2 \cdot 5 && \text{Commutative Property} \\
&= (6 \cdot 3)(2 \cdot 5) && \text{Associative Property} \\
&= 18 \cdot 10 && \text{Multiply.} \\
&= 180 && \text{Multiply.}
\end{aligned}
$$

The product is 180.

Example 2   Evaluate $8.2 + 2.5 + 2.5 + 1.8$ using properties of numbers. Name the property used in each step.

$$
\begin{aligned}
8.2 &+ 2.5 + 2.5 + 1.8 \\
&= 8.2 + 1.8 + 2.5 + 2.5 && \text{Commutative Prop.} \\
&= (8.2 + 1.8) + (2.5 + 2.5) && \text{Associative Prop.} \\
&= 10 + 5 && \text{Add.} \\
&= 15 && \text{Add.}
\end{aligned}
$$

The sum is 15.

Exercises

Evaluate each expression using properties of numbers. Name the property used in each step.

1. $12 + 10 + 8 + 5$

2. $16 + 8 + 22 + 12$

3. $10 \cdot 7 \cdot 2.5$

4. $4 \cdot 8 \cdot 5 \cdot 3$

5. $12 + 20 + 10 + 5$

6. $26 + 8 + 4 + 22$

7. $3\frac{1}{2} + 4 + 2\frac{1}{2} + 3$

8. $\frac{3}{4} \cdot 12 \cdot 4 \cdot 2$

9. $3.5 + 2.4 + 3.6 + 4.2$

10. $4\frac{1}{2} + 5 + \frac{1}{2} + 3$

11. $0.5 \cdot 2.8 \cdot 4$

12. $2.5 + 2.4 + 2.5 + 3.6$

13. $\frac{4}{5} \cdot 18 \cdot 25 \cdot \frac{2}{9}$

14. $32 \cdot \frac{1}{5} \cdot \frac{1}{2} \cdot 10$

15. $\frac{1}{4} \cdot 7 \cdot 16 \cdot \frac{1}{7}$

16. $3.5 + 8 + 2.5 + 2$

17. $18 \cdot 8 \cdot \frac{1}{2} \cdot \frac{1}{9}$

18. $\frac{3}{4} \cdot 10 \cdot 16 \cdot \frac{1}{2}$

1-4 Study Guide and Intervention

The Distributive Property

Evaluate Expressions The Distributive Property can be used to help evaluate expressions.

Distributive Property	For any numbers a, b, and c, $a(b + c) = ab + ac$ and $(b + c)a = ba + ca$ and $a(b - c) = ab - ac$ and $(b - c)a = ba - ca$.

Example 1 Use the Distributive Property to rewrite $6(8 + 10)$. Then evaluate.

$$
\begin{aligned}
6(8 + 10) &= 6 \cdot 8 + 6 \cdot 10 &&\text{Distributive Property}\\
&= 48 + 60 &&\text{Multiply.}\\
&= 108 &&\text{Add.}
\end{aligned}
$$

Example 2 Use the Distributive Property to rewrite $-2(3x^2 + 5x + 1)$. Then simplify.

$$
\begin{aligned}
-2(3x^2 + 5x + 1) &= -2(3x^2) + (-2)(5x) + (-2)(1) &&\text{Distributive Property}\\
&= -6x^2 + (-10x) + (-2) &&\text{Multiply.}\\
&= -6x^2 - 10x - 2 &&\text{Simplify.}
\end{aligned}
$$

Exercises

Use the Distributive Property to rewrite each expression. Then evaluate.

1. $20(31)$

2. $12 \cdot 4\frac{1}{2}$

3. $5(311)$

4. $5(4x - 9)$

5. $3(8 - 2x)$

6. $12\left(6 - \frac{1}{2}x\right)$

7. $12\left(2 + \frac{1}{2}x\right)$

8. $\frac{1}{4}(12 - 4t)$

9. $3(2x - y)$

10. $2(3x + 2y - z)$

11. $(x - 2)y$

12. $2(3a - 2b + c)$

13. $\frac{1}{4}(16x - 12y + 4z)$

14. $(2 - 3x + x^2)3$

15. $-2(2x^2 + 3x + 1)$

1-4 Study Guide and Intervention (continued)

The Distributive Property

Simplify Expressions A **term** is a number, a variable, or a product or quotient of numbers and variables. **Like terms** are terms that contain the same variables, with corresponding variables having the same powers. The Distributive Property and properties of equalities can be used to simplify expressions. An expression is in **simplest form** if it is replaced by an **equivalent** expression with no like terms or parentheses.

Example Simplify $4(a^2 + 3ab) - ab$.

$$
\begin{aligned}
4(a^2 + 3ab) - ab &= 4(a^2 + 3ab) - 1ab && \text{Multiplicative Identity} \\
&= 4a^2 + 12ab - 1ab && \text{Distributive Property} \\
&= 4a^2 + (12 - 1)ab && \text{Distributive Property} \\
&= 4a^2 + 11ab && \text{Substitution}
\end{aligned}
$$

Exercises

Simplify each expression. If not possible, write *simplified*.

1. $12a - a$

2. $3x + 6x$

3. $3x - 1$

4. $20a + 12a - 8$

5. $3x^2 + 2x^2$

6. $-6x + 3x^2 + 10x^2$

7. $2p + \frac{1}{2}q$

8. $10xy - 4(xy + xy)$

9. $21a + 18a + 31b - 3b$

10. $4x + \frac{1}{4}(16x - 20y)$

11. $2 - 1 - 6x + x^2$

12. $4x^2 + 3x^2 + 2x$

Write an algebraic expression for each verbal expression. Then simplify, indicating the properties used.

13. six times the difference of $2a$ and b, increased by $4b$

14. two times the sum of x squared and y squared, increased by three times the sum of x squared and y squared

 Glencoe Algebra 1

1-5 Study Guide and Intervention

Equations

Solve Equations A mathematical sentence with one or more variables is called an **open sentence**. Open sentences are **solved** by finding replacements for the variables that result in true sentences. The set of numbers from which replacements for a variable may be chosen is called the **replacement set**. The set of all replacements for the variable that result in true statements is called the **solution set** for the variable. A sentence that contains an equal sign, =, is called an **equation**.

Example 1 Find the solution set of $3a + 12 = 39$ if the replacement set is {6, 7, 8, 9, 10}.

Replace a in $3a + 12 = 39$ with each value in the replacement set.

$3(6) + 12 \stackrel{?}{=} 39 \rightarrow 30 \neq 39$ false
$3(7) + 12 \stackrel{?}{=} 39 \rightarrow 33 \neq 39$ false
$3(8) + 12 \stackrel{?}{=} 39 \rightarrow 36 \neq 39$ false
$3(9) + 12 \stackrel{?}{=} 39 \rightarrow 39 = 39$ true
$3(10) + 12 \stackrel{?}{=} 39 \rightarrow 42 \neq 39$ false

Since $a = 9$ makes the equation $3a + 12 = 39$ true, the solution is 9.

The solution set is {9}.

Example 2 Solve $\dfrac{2(3 + 1)}{3(7 - 4)} = b$.

$\dfrac{2(3 + 1)}{3(7 - 4)} = b$ Original equation

$\dfrac{2(4)}{3(3)} = b$ Add in the numerator; subtract in the denominator.

$\dfrac{8}{9} = b$ Simplify.

The solution is $\dfrac{8}{9}$.

Exercises

Find the solution of each equation if the replacement sets are $x = \left\{\dfrac{1}{4}, \dfrac{1}{2}, 1, 2, 3\right\}$ and $y = \{2, 4, 6, 8\}$.

1. $x + \dfrac{1}{2} = \dfrac{5}{2}$

2. $x + 8 = 11$

3. $y - 2 = 6$

4. $x^2 - 1 = 8$

5. $y^2 - 2 = 34$

6. $x^2 + 5 = 5\dfrac{1}{16}$

7. $2(x + 3) = 7$

8. $(y + 1)^2 = 9$

9. $y^2 + y = 20$

Solve each equation.

10. $a = 2^3 - 1$

11. $n = 6^2 - 4^2$

12. $w = 6^2 \cdot 3^2$

13. $\dfrac{1}{4} + \dfrac{5}{8} = k$

14. $\dfrac{18 - 3}{2 + 3} = p$

15. $t = \dfrac{15 - 6}{27 - 24}$

16. $18.4 - 3.2 = m$

17. $k = 9.8 + 5.7$

18. $c = 3\dfrac{1}{2} + 2\dfrac{1}{4}$

1-5   Study Guide and Intervention (continued)

Equations

Solve Equations with Two Variables Some equations contain two variables. It is often useful to make a table of values in which you can use substitution to find the corresponding values of the second variable.

MUSIC DOWNLOADS Emily belongs to an Internet music service that charges $5.99 per month and $0.89 per song. Write and solve an equation to find the total amount Emily spends if she downloads 10 songs this month.

The cost of the music service is a flat rate. The variable is the number of songs she downloads. The total cost is the price of the service plus $0.89 times the number of songs.

$C = 0.89n + 5.99$

To find the total cost for the month, substitute 10 for n in the equation.

$C = 0.89n + 5.99$	Original equation
$= 0.89(10) + 5.99$	Substitute 10 for n.
$= 8.90 + 5.99$	Multiply.
$= 14.89$	Add.

Emily spent $14.89 on music downloads in one month.

Exercises

1. **AUTO REPAIR** A mechanic repairs Mr. Estes' car. The amount for parts is $48.00 and the rate for the mechanic is $40.00 per hour. Write and solve an equation to find the total cost of repairs to Mr. Estes' car if the mechanic works for 1.5 hours.

2. **SHIPPING FEES** Mr. Moore purchases an inflatable kayak weighing 30 pounds from an online company. The standard rate to ship his purchase is $2.99 plus $0.85 per pound. Write and solve an equation to find the total amount Mr. Moore will pay to have the kayak shipped to his home.

3. **SOUND** The speed of sound is 1088 feet per second at sea level at 32°F. Write and solve an equation to find the distance sound travels in 8 seconds under these conditions.

4. **VOLLEYBALL** Your town decides to build a volleyball court. If the court is approximately 40 by 70 feet and its surface is of sand, one foot deep, the court will require about 166 tons of sand. A local sand pit sells sand for $11.00 per ton with a delivery charge of $3.00 per ton. Write and solve an equation to find the total cost of the sand for this court.

1-6 Study Guide and Intervention

Relations

Represent a Relation A **relation** is a set of ordered pairs. A relation can be represented by a set of ordered pairs, a table, a graph, or a **mapping**. A mapping illustrates how each element of the domain is paired with an element in the range. The set of first numbers of the ordered pairs is the **domain**. The set of second numbers of the ordered pairs is the **range** of the relation.

Example   **a. Express the relation {(1, 1), (0, 2), (3, −2)} as a table, a graph, and a mapping.**

x	y
1	1
0	2
3	−2

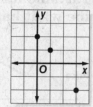

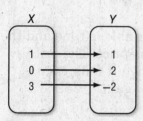

b. Determine the domain and the range of the relation.

The domain for this relation is {0, 1, 3}. The range for this relation is {−2, 1, 2}.

Exercises

1A. Express the relation {(−2, −1), (3, 3), (4, 3)} as a table, a graph, and a mapping.

x	y

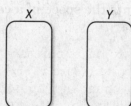

1B. Determine the domain and the range of the relation.

1-6 Study Guide and Intervention (continued)

Relations

Graphs of a Relation The value of the variable in a relation that is subject to choice is called the **independent variable**. The variable with a value that is dependent on the value of the independent variable is called the **dependent variable**. These relations can be graphed without a scale on either axis, and interpreted by analyzing the shape.

Example 1 The graph below represents the height of a football after it is kicked downfield. Identify the independent and the dependent variable for the relation. Then describe what happens in the graph.

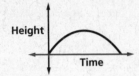

The independent variable is time, and the dependent variable is height. The football starts on the ground when it is kicked. It gains altitude until it reaches a maximum height, then it loses altitude until it falls to the ground.

Example 2 The graph below represents the price of stock over time. Identify the independent and dependent variable for the relation. Then describe what happens in the graph.

The independent variable is time and the dependent variable is price. The price increases steadily, then it falls, then increases, then falls again.

Exercises

Identify the independent and dependent variables for each relation. Then describe what is happening in each graph.

1. The graph represents the speed of a car as it travels to the grocery store.

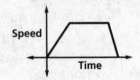

2. The graph represents the balance of a savings account over time.

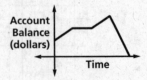

3. The graph represents the height of a baseball after it is hit.

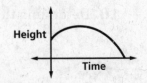

 Copyright © Glencoe/McGraw-Hill, a division of The McGraw-Hill Companies, Inc.

Glencoe Algebra 1

1-7 Study Guide and Intervention
Functions

Identify Functions Relations in which each element of the domain is paired with exactly one element of the range are called **functions**.

Example 1 Determine whether the relation {(6, −3), (4, 1), (7, −2), (−3, 1)} is a function. Explain.

Since each element of the domain is paired with exactly one element of the range, this relation is a function.

Example 2 Determine whether $3x - y = 6$ is a function.

Since the equation is in the form $Ax + By = C$, the graph of the equation will be a line, as shown at the right.

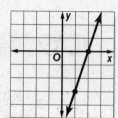

If you draw a vertical line through each value of x, the vertical line passes through just one point of the graph. Thus, the line represents a function.

Exercises

Determine whether each relation is a function.

1.

2.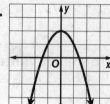

3.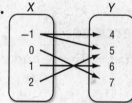

4.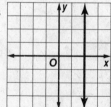

5.

6.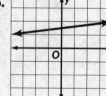

7. {(4, 2), (2, 3), (6, 1)} **8.** {(−3, −3), (−3, 4), (−2, 4)} **9.** {(−1, 0), (1, 0)}

10. $-2x + 4y = 0$ **11.** $x^2 + y^2 = 8$ **12.** $x = -4$

1-7 Study Guide and Intervention (continued)

Functions

Find Function Values Equations that are functions can be written in a form called **function notation**. For example, $y = 2x - 1$ can be written as $f(x) = 2x - 1$. In the function, x represents the elements of the domain, and $f(x)$ represents the elements of the range. Suppose you want to find the value in the range that corresponds to the element 2 in the domain. This is written $f(2)$ and is read "f of 2." The value of $f(2)$ is found by substituting 2 for x in the equation.

Example If $f(x) = 3x - 4$, find each value.

a. $f(3)$

$f(3) = 3(3) - 4$	Replace x with 3.
$= 9 - 4$	Multiply.
$= 5$	Simplify.

b. $f(-2)$

$f(-2) = 3(-2) - 4$	Replace x with -2.
$= -6 - 4$	Multiply.
$= -10$	Simplify.

Exercises

If $f(x) = 2x - 4$ and $g(x) = x^2 - 4x$, find each value.

1. $f(4)$ **2.** $g(2)$ **3.** $f(-5)$

4. $g(-3)$ **5.** $f(0)$ **6.** $g(0)$

7. $f(3) - 1$ **8.** $f\left(\dfrac{1}{4}\right)$ **9.** $g\left(\dfrac{1}{4}\right)$

10. $f(a^2)$ **11.** $f(k + 1)$ **12.** $g(2n)$

13. $f(3x)$ **14.** $f(2) + 3$ **15.** $g(-4)$

1-8 Study Guide and Intervention

Interpreting Graphs of Functions

Interpret Intercepts and Symmetry The **intercepts** of a graph are points where the graph intersects an axis. The y-coordinate of the point at which the graph intersects the y-axis is called a **y-intercept.** Similarly, the x-coordinate of the point at which a graph intersects the x-axis is called an **x-intercept.**

A graph possesses **line symmetry** in a line if each half of the graph on either side of the line matches exactly.

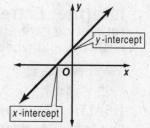

Example

ARCHITECTURE The graph shows a function that approximates the shape of the Gateway Arch, where x is the distance from the center point in feet and y is the height in feet. Identify the function as *linear* or *nonlinear*. Then estimate and interpret the intercepts, and describe and interpret any symmetry.

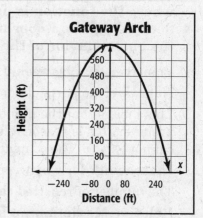

Linear or Nonlinear: Since the graph is a curve and not a line, the graph is nonlinear.

y-Intercept: The graph intersects the y-axis at about (0, 630), so the y-intercept of the graph is about 630. This means that the height of the arch is 630 feet at the center point.

x-Intercept(s): The graph intersects the x-axis at about (−320, 0) and (320, 0). So the x-intercepts are about −320 and 320. This means that the object touches the ground to the left and right of the center point.

Symmetry: The right half of the graph is the mirror image of the left half in the y-axis. In the context of the situation, the symmetry of the graph tells you that the arch is symmetric. The height of the arch at any distance to the right of the center is the same as its height that same distance to the left.

Identify the function graphed as *linear* or *nonlinear*. Then estimate and interpret the intercepts of the graph and any symmetry.

1.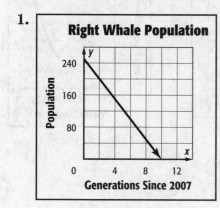

2.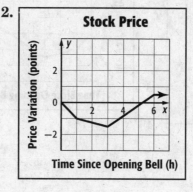

3.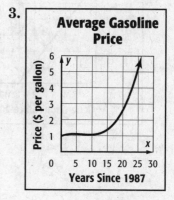

1-8 Study Guide and Intervention *(continued)*

Interpreting Graphs of Functions

Interpret Extrema and End Behavior Interpreting a graph also involves estimating and interpreting where the function is increasing, decreasing, positive, or negative, and where the function has any extreme values, either high or low.

> **Example**

HEALTH The outbreak of the H1N1 virus can be modeled by the function graphed at the right. Estimate and interpret where the function is positive, negative, increasing, and decreasing, the *x*-coordinates of any relative extrema, and the end behavior of the graph.

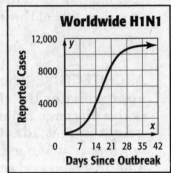

Worldwide H1N1

Positive: for *x* between 0 and 42

Negative: no parts of domain

This means that the number of reported cases was always positive. This is reasonable because a negative number of cases cannot exist in the context of the situation.

Increasing: for *x* between 0 and 42   **Decreasing:** no parts of domain

The number of reported cases increased each day from the first day of the outbreak.

Relative Maximum: at about $x = 42$ **Relative Minimum:** at $x = 0$

The extrema of the graph indicate that the number of reported cases peaked at about day 42.

End Behavior: As *x* increases, *y* appears to approach 11,000. As *x* decreases, *y* decreases. The end behavior of the graph indicates a maximum number of reported cases of 11,000.

Estimate and interpret where the function is positive, negative, increasing, and decreasing, the *x*-coordinate of any relative extrema, and the end behavior of the graph.

1.

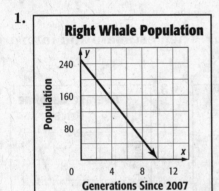

Right Whale Population

2.

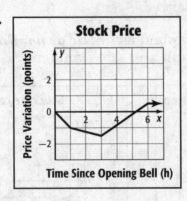

Stock Price

3.

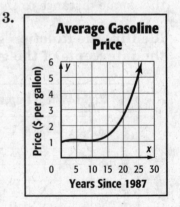

Average Gasoline Price

2-1 Study Guide and Intervention

Writing Equations

Write Equations Writing equations is one strategy for solving problems. You can use a variable to represent an unspecified number or measure referred to in a problem. Then you can write a verbal expression as an algebraic expression.

Example 1 Translate each sentence into an equation or a formula.

a. **Ten times a number x is equal to 2.8 times the difference y minus z.**

$10 \times x = 2.8 \times (y - z)$
The equation is $10x = 2.8(y - z)$.

b. **A number m minus 8 is the same as a number n divided by 2.**

$m - 8 = n \div 2$
The equation is $m - 8 = \frac{n}{2}$.

c. **The area of a rectangle equals the length times the width. Translate this sentence into a formula.**

Let A = area, ℓ = length, and w = width.
Formula: *Area equals length times width.*
$A = \ell \times w$
The formula for the area of a rectangle is $A = \ell w$.

Example 2 Use the Four-Step Problem-Solving Plan.

POPULATION The population of the United States in July 2007 was about 301,000,000, and the land area of the United States is about 3,500,000 square miles. Find the average number of people per square mile in the United States.

Step 1 *Read* You know that there are 301,000,000 people. You want to know the number of people per square mile.

Step 2 *Plan* Write an equation to represent the situation. Let p represent the number of people per square mile.
$3{,}500{,}000 \times p = 301{,}000{,}000$

Step 3 *Solve* $3{,}500{,}000 \times p = 301{,}000{,}000$.
$3{,}500{,}000p = 301{,}000{,}000$ Divide each side by
$p = 86$ 3,500,000.
There are 86 people per square mile.

Step 4 *Check* If there are 86 people per square mile and there are 3,500,000 square miles, $86 \times 3{,}500{,}000 = 301{,}000{,}000$. The answer makes sense.

Exercises

Translate each sentence into an equation or formula.

1. Three times a number t minus twelve equals forty.

2. One-half of the difference of a and b is 54.

3. Three times the sum of d and 4 is 32.

4. The area A of a circle is the product of π and the radius r squared.

5. **WEIGHT LOSS** Lou wants to lose weight to audition for a part in a play. He weighs 160 pounds now. He wants to weigh 150 pounds.

   a. If p represents the number of pounds he wants to lose, write an equation to represent this situation.

   b. How many pounds does he need to lose to reach his goal?

2-1 Study Guide and Intervention (continued)

Writing Equations

Write Verbal Sentences You can translate equations into verbal sentences.

| Example | Translate each equation into a sentence.

a. $4n - 8 = 12$.

$$4n \quad\quad - \quad 8 \quad = \quad 12$$

Four times n minus eight equals twelve.

b. $a^2 + b^2 = c^2$

$$a^2 + b^2 \quad\quad\quad\quad = \quad\quad c^2$$

The sum of the squares of a and b is equal to the square of c.

Exercises

Translate each equation into a sentence.

1. $4a - 5 = 23$

2. $10 + k = 4k$

3. $6xy = 24$

4. $x^2 + y^2 = 8$

5. $p + 3 = 2p$

6. $b = \frac{1}{3}(h - 1)$

7. $100 - 2x = 80$

8. $3(g + h) = 12$

9. $p^2 - 2p = 9$

10. $C = \frac{5}{9}(F - 32)$

11. $V = \frac{1}{3}Bh$

12. $A = \frac{1}{2}hb$

2-2 Study Guide and Intervention

Solving One-Step Equations

Solve Equations Using Addition and Subtraction If the same number is added to each side of an equation, the resulting equation is equivalent to the original one. In general if the original equation involves subtraction, this property will help you solve the equation. Similarly, if the same number is subtracted from each side of an equation, the resulting equation is equivalent to the original one. This property will help you solve equations involving addition.

| Addition Property of Equality | For any numbers a, b, and c, if $a = b$, then $a + c = b + c$. |
| Subtraction Property of Equality | For any numbers a, b, and c, if $a = b$, then $a - c = b - c$. |

Example 1 Solve $m - 32 = 18$.

$$m - 32 = 18 \quad \text{Original equation}$$
$$m - 32 + 32 = 18 + 32 \quad \text{Add 32 to each side.}$$
$$m = 50 \quad \text{Simplify.}$$

The solution is 50.

Example 2 Solve $22 + p = -12$.

$$22 + p = -12 \quad \text{Original equation}$$
$$22 + p - 22 = -12 - 22 \quad \text{Subtract 22 from each side.}$$
$$p = -34 \quad \text{Simplify.}$$

The solution is -34.

Exercises

Solve each equation. Check your solution.

1. $h - 3 = -2$

2. $m - 8 = -12$

3. $p - 5 = 15$

4. $20 = y - 8$

5. $k - 0.5 = 2.3$

6. $w - \dfrac{1}{2} = \dfrac{5}{8}$

7. $h - 18 = -17$

8. $-12 = -24 + k$

9. $j - 0.2 = 1.8$

10. $b - 40 = -40$

11. $m - (-12) = 10$

12. $w - \dfrac{3}{2} = \dfrac{1}{4}$

13. $x + 12 = 6$

14. $w + 2 = -13$

15. $-17 = b + 4$

16. $k + (-9) = 7$

17. $-3.2 = \ell + (-0.2)$

18. $-\dfrac{3}{8} + x = \dfrac{5}{8}$

19. $19 + h = -4$

20. $-12 = k + 24$

21. $j + 1.2 = 2.8$

22. $b + 80 = -80$

23. $m + (-8) = 2$

24. $w + \dfrac{3}{2} = \dfrac{5}{8}$

2-2 Study Guide and Intervention (continued)

Solving One-Step Equations

Solve Equations Using Multiplication and Division If each side of an equation is multiplied by the same number, the resulting equation is equivalent to the given one. You can use the property to solve equations involving multiplication and division. To solve equations with multiplication and division, you can also use the Division Property of Equality. If each side of an equation is divided by the same number, the resulting equation is true.

| Multiplication Property of Equality | For any numbers a, b, and c, if $a = b$, then $ac = bc$. |
| Division Property of Equality | For any numbers a, b, and c, with $c \neq 0$, if $a = b$, then $\frac{a}{c} = \frac{b}{c}$. |

Example 1 Solve $3\frac{1}{2}p = 1\frac{1}{2}$.

$3\frac{1}{2}p = 1\frac{1}{2}$	Original equation
$\frac{7}{2}p = \frac{3}{2}$	Rewrite each mixed number as an improper fraction.
$\frac{2}{7}\left(\frac{7}{2}p\right) = \frac{2}{7}\left(\frac{3}{2}\right)$	Multiply each side by $\frac{2}{7}$.
$p = \frac{3}{7}$	Simplify.

The solution is $\frac{3}{7}$.

Example 2 Solve $-5n = 60$.

$-5n = 60$	Original equation
$\frac{-5n}{-5} = \frac{60}{-5}$	Divide each side by -5.
$n = -12$	Simplify.

The solution is -12.

Exercises

Solve each equation. Check your solution.

1. $\frac{h}{3} = -2$ **2.** $\frac{1}{8}m = 6$ **3.** $\frac{1}{5}p = \frac{3}{5}$

4. $5 = \frac{y}{12}$ **5.** $-\frac{1}{4}k = -2.5$ **6.** $-\frac{m}{8} = \frac{5}{8}$

7. $-1\frac{1}{2}h = 4$ **8.** $-12 = -\frac{3}{2}k$ **9.** $\frac{j}{3} = \frac{2}{5}$

10. $-3\frac{1}{3}b = 5$ **11.** $\frac{7}{10}m = 10$ **12.** $\frac{p}{5} = -\frac{1}{4}$

13. $3h = -42$ **14.** $8m = 16$ **15.** $-3t = 51$

16. $-3r = -24$ **17.** $8k = -64$ **18.** $-2m = 16$

19. $12h = 4$ **20.** $-2.4p = 7.2$ **21.** $0.5j = 5$

22. $-25 = 5m$ **23.** $6m = 15$ **24.** $-1.5p = -75$

2-3 Study Guide and Intervention
Solving Multi-Step Equations

Work Backward **Working backward** is one of many problem-solving strategies that you can use to solve problems. To work backward, start with the result given at the end of a problem and undo each step to arrive at the beginning number.

Example 1 **A number is divided by 2, and then 8 is subtracted from the quotient. The result is 16. What is the number?**

Solve the problem by working backward.

The final number is 16. Undo subtracting 8 by adding 8 to get 24. To undo dividing 24 by 2, multiply 24 by 2 to get 48.

The original number is 48.

Example 2 **A bacteria culture doubles each half hour. After 3 hours, there are 6400 bacteria. How many bacteria were there to begin with?**

Solve the problem by working backward.

The bacteria have grown for 3 hours. Since there are 2 one-half hour periods in one hour, in 3 hours there are 6 one-half hour periods. Since the bacteria culture has grown for 6 time periods, it has doubled 6 times. Undo the doubling by halving the number of bacteria 6 times.

$$6400 \times \frac{1}{2} \times \frac{1}{2} \times \frac{1}{2} \times \frac{1}{2} \times \frac{1}{2} \times \frac{1}{2} = 6400 \times \frac{1}{64}$$
$$= 100$$

There were 100 bacteria to begin with.

Exercises

Solve each problem by working backward.

1. A number is divided by 3, and then 4 is added to the quotient. The result is 8. Find the number.

2. A number is multiplied by 5, and then 3 is subtracted from the product. The result is 12. Find the number.

3. Eight is subtracted from a number, and then the difference is multiplied by 2. The result is 24. Find the number.

4. Three times a number plus 3 is 24. Find the number.

5. **CAR RENTAL** Angela rented a car for $29.99 a day plus a one-time insurance cost of $5.00. Her bill was $124.96. For how many days did she rent the car?

6. **MONEY** Mike withdrew an amount of money from his bank account. He spent one fourth for gasoline and had $90 left. How much money did he withdraw?

2-3 Study Guide and Intervention (continued)

Solving Multi-Step Equations

Solve Multi-Step Equations To solve equations with more than one operation, often called **multi-step equations**, undo operations by working backward. Reverse the usual order of operations as you work.

Example	Solve $5x + 3 = 23$.

$5x + 3 = 23$	Original equation
$5x + 3 - 3 = 23 - 3$	Subtract 3 from each side.
$5x = 20$	Simplify.
$\dfrac{5x}{5} = \dfrac{20}{5}$	Divide each side by 5.
$x = 4$	Simplify.

Exercises

Solve each equation. Check your solution.

1. $5x + 2 = 27$

2. $6x + 9 = 27$

3. $5x + 16 = 51$

4. $14n - 8 = 34$

5. $0.6x - 1.5 = 1.8$

6. $\dfrac{7}{8}p - 4 = 10$

7. $16 = \dfrac{d - 12}{14}$

8. $8 + \dfrac{3n}{12} = 13$

9. $\dfrac{g}{-5} + 3 = -13$

10. $\dfrac{4b + 8}{-2} = 10$

11. $0.2x - 8 = -2$

12. $3.2y - 1.8 = 3$

13. $-4 = \dfrac{7x - (-1)}{-8}$

14. $8 = -12 + \dfrac{k}{-4}$

15. $0 = 10y - 40$

Write an equation and solve each problem.

16. Find three consecutive integers whose sum is 96.

17. Find two consecutive odd integers whose sum is 176.

18. Find three consecutive integers whose sum is -93.

2-4 Study Guide and Intervention

Solving Equations with the Variable on Each Side

Variables on Each Side To solve an equation with the same variable on each side, first use the Addition or the Subtraction Property of Equality to write an equivalent equation that has the variable on just one side of the equation. Then solve the equation.

Example 1 Solve $5y - 8 = 3y + 12$.

$$5y - 8 = 3y + 12$$
$$5y - 8 - 3y = 3y + 12 - 3y$$
$$2y - 8 = 12$$
$$2y - 8 + 8 = 12 + 8$$
$$2y = 20$$
$$\frac{2y}{2} = \frac{20}{2}$$
$$y = 10$$

The solution is 10.

Example 2 Solve $-11 - 3y = 8y + 1$.

$$-11 - 3y = 8y + 1$$
$$-11 - 3y + 3y = 8y + 1 + 3y$$
$$-11 = 11y + 1$$
$$-11 - 1 = 11y + 1 - 1$$
$$-12 = 11y$$
$$\frac{-12}{11} = \frac{11y}{11}$$
$$-1\frac{1}{11} = y$$

The solution is $-1\frac{1}{11}$.

Exercises

Solve each equation. Check your solution.

1. $6 - b = 5b + 30$

2. $5y - 2y = 3y + 2$

3. $5x + 2 = 2x - 10$

4. $4n - 8 = 3n + 2$

5. $1.2x + 4.3 = 2.1 - x$

6. $4.4m + 6.2 = 8.8m - 1.8$

7. $\frac{1}{2}b + 4 = \frac{1}{8}b + 88$

8. $\frac{3}{4}k - 5 = \frac{1}{4}k - 1$

9. $8 - 5p = 4p - 1$

10. $4b - 8 = 10 - 2b$

11. $0.2x - 8 = -2 - x$

12. $3y - 1.8 = 3y - 1.8$

13. $-4 - 3x = 7x - 6$

14. $8 + 4k = -10 + k$

15. $20 - a = 10a - 2$

16. $\frac{2}{3}n + 8 = \frac{1}{2}n + 2$

17. $\frac{2}{5}y - 8 = 9 - \frac{3}{5}y$

18. $-4r + 5 = 5 - 4r$

19. $-4 - 3x = 6x - 6$

20. $18 - 4k = -10 - 4k$

21. $12 + 2y = 10y - 12$

2-4 Study Guide and Intervention (continued)

Solving Equations with the Variable on Each Side

Grouping Symbols When solving equations that contain grouping symbols, first use the Distributive Property to eliminate grouping symbols. Then solve.

Example Solve $4(2a - 1) = -10(a - 5)$.

$4(2a - 1) = -10(a - 5)$	Original equation
$8a - 4 = -10a + 50$	Distributive Property
$8a - 4 + 10a = -10a + 50 + 10a$	Add 10a to each side.
$18a - 4 = 50$	Simplify.
$18a - 4 + 4 = 50 + 4$	Add 4 to each side.
$18a = 54$	Simplify.
$\dfrac{18a}{18} = \dfrac{54}{18}$	Divide each side by 18.
$a = 3$	Simplify.

The solution is 3.

Exercises

Solve each equation. Check your solution.

1. $-3(x + 5) = 3(x - 1)$

2. $2(7 + 3t) = -t$

3. $3(a + 1) - 5 = 3a - 2$

4. $75 - 9g = 5(-4 + 2g)$

5. $5(f + 2) = 2(3 - f)$

6. $4(p + 3) = 36$

7. $18 = 3(2t + 2)$

8. $3(d - 8) = 3d$

9. $5(p + 3) + 9 = 3(p - 2) + 6$

10. $4(b - 2) = 2(5 - b)$

11. $1.2(x - 2) = 2 - x$

12. $\dfrac{3 + y}{4} = \dfrac{-y}{8}$

13. $\dfrac{a - 8}{12} = \dfrac{2a + 5}{3}$

14. $2(4 + 2k) + 10 = k$

15. $2(w - 1) + 4 = 4(w + 1)$

16. $6(n - 1) = 2(2n + 4)$

17. $2[2 + 3(y - 1)] = 22$

18. $-4(r + 2) = 4(2 - 4r)$

19. $-3(x - 8) = 24$

20. $4(4 - 4k) = -10 - 16k$

21. $6(2 - 2y) = 5(2y - 2)$

2-5 Study Guide and Intervention
Solving Equations Involving Absolute Value

Absolute Value Expressions Expressions with absolute values define an upper and lower range in which a value must lie. Expressions involving absolute value can be evaluated using the given value for the variable.

Example	**Evaluate $|t - 5| - 7$ if $t = 3$.**

$|t - 5| - 7 = |3 - 5| - 7$ Replace t with 3.

$\qquad\qquad = |-2| - 7$ $3 - 5 = -2$

$\qquad\qquad = 2 - 7$ $|-2| = 2$

$\qquad\qquad = -5$ Simplify.

Exercises

Evaluate each expression if $r = -2$, $n = -3$, and $t = 3$.

1. $|8 - t| + 3$ **2.** $|t - 3| - 7$ **3.** $5 + |3 - n|$

4. $|r + n| - 7$ **5.** $|n - t| + 4$ **6.** $-|r + n + t|$

Evaluate each expression if $n = 2$, $q = -1.5$, $r = -3$, $v = -8$, $w = 4.5$, and $x = 4$.

7. $|2q + r|$ **8.** $10 - |2n + v|$ **9.** $|3x - 2w| - q$

10. $v - |3n + x|$ **11.** $1 + |5q - w|$ **12.** $2|3r - v|$

13. $|-2x + 5n| + (n - x)$ **14.** $4w - |2r + v|$ **15.** $3|w - n| - 5|q - r|$

2-5 **Study Guide and Intervention** (continued)

Solving Equations Involving Absolute Value

Absolute Value Equations When solving equations that involve absolute value, there are two cases to consider.

Case 1: The value inside the absolute value symbols is positive.
Case 2: The value inside the absolute value symbols is negative.

Example 1 Solve $|x + 4| = 1$. Then graph the solution set.

Write $|x + 4| = 1$ as $x + 4 = 1$ or $x + 4 = -1$.

$$x + 4 = 1 \qquad \text{or} \qquad x + 4 = -1$$
$$x + 4 - 4 = 1 - 4 \qquad x + 4 - 4 = -1 - 4$$
$$x = -3 \qquad\qquad x = -5$$

The solution set is $\{-5, -3\}$.
The graph is shown below.

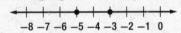

Example 2 Write an equation involving absolute value for the graph.

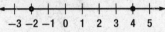

Find the point that is the same distance from -2 as it is from 4.

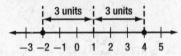

The distance from 1 to -2 is 3 units. The distance from 1 to 4 is 3 units.
So, $|x - 1| = 3$.

Exercises

Solve each equation. Then graph the solution set.

1. $|y| = 3$

2. $|x - 4| = 4$

3. $|y + 3| = 2$

4. $|b + 2| = 3$

5. $|w - 2| = 5$

6. $|t + 2| = 4$

7. $|2x| = 8$

8. $|5y - 2| = 7$

9. $|p - 0.2| = 0.5$

10. $|d - 100| = 50$

11. $|2x - 1| = 11$

12. $\left|3x + \dfrac{1}{2}\right| = 6$

Write an equation involving absolute value for each graph.

13.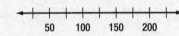

14.

15.

2-6 Study Guide and Intervention

Ratios and Proportions

Ratios and Proportions A **ratio** is a comparison of two numbers by division. The ratio of x to y can be expressed as x to y, $x{:}y$ or $\frac{x}{y}$. Ratios are usually expressed in simplest form. An equation stating that two ratios are equal is called a **proportion**. To determine whether two ratios form a proportion, express both ratios in simplest form or check cross products.

Example 1	Example 2
Determine whether the ratios $\frac{24}{36}$ and $\frac{12}{18}$ are equivalent ratios. Write *yes* or *no*. Justify your answer.	**Use cross products to determine whether $\frac{10}{18}$ and $\frac{25}{45}$ form a proportion.**

Example 1 Determine whether the ratios $\frac{24}{36}$ and $\frac{12}{18}$ are equivalent ratios. Write *yes* or *no*. Justify your answer.

$\frac{24}{36} = \frac{2}{3}$ when expressed in simplest form.

$\frac{12}{18} = \frac{2}{3}$ when expressed in simplest form.

The ratios $\frac{24}{36}$ and $\frac{12}{18}$ form a proportion because they are equal when expressed in simplest form.

Example 2 Use cross products to determine whether $\frac{10}{18}$ and $\frac{25}{45}$ form a proportion.

$\frac{10}{18} \overset{?}{=} \frac{25}{45}$ Write the proportion.

$10(45) \overset{?}{=} 18(25)$ Cross products

$450 = 450$ Simplify.

The cross products are equal, so $\frac{10}{18} = \frac{25}{45}$. Since the ratios are equal, they form a proportion.

Exercises

Determine whether each pair of ratios are equivalent ratios. Write *yes* or *no*.

1. $\frac{1}{2}, \frac{16}{32}$

2. $\frac{5}{8}, \frac{10}{15}$

3. $\frac{10}{20}, \frac{25}{49}$

4. $\frac{25}{36}, \frac{15}{20}$

5. $\frac{12}{32}, \frac{3}{16}$

6. $\frac{4}{9}, \frac{12}{27}$

7. $\frac{0.1}{2}, \frac{5}{100}$

8. $\frac{15}{20}, \frac{9}{12}$

9. $\frac{14}{12}, \frac{20}{30}$

10. $\frac{2}{3}, \frac{20}{30}$

11. $\frac{5}{9}, \frac{25}{45}$

12. $\frac{72}{64}, \frac{9}{8}$

13. $\frac{5}{5}, \frac{30}{20}$

14. $\frac{18}{24}, \frac{50}{75}$

15. $\frac{100}{75}, \frac{44}{33}$

16. $\frac{0.05}{1}, \frac{1}{20}$

17. $\frac{1.5}{2}, \frac{6}{8}$

18. $\frac{0.1}{0.2}, \frac{0.45}{0.9}$

2-6 Study Guide and Intervention (continued)

Ratios and Proportions

Solve Proportions If a proportion involves a variable, you can use cross products to solve the proportion. In the proportion $\frac{x}{5} = \frac{10}{13}$, x and 13 are called **extremes**. They are the first and last terms of the proportion. 5 and 10 are called **means**. They are the middle terms of the proportion. In a proportion, the product of the extremes is equal to the product of the means.

Means-Extremes Property of Proportions	For any numbers a, b, c, and d, if $\frac{a}{b} = \frac{c}{d}$, then $ad = bc$.

Example Solve $\frac{x}{5} = \frac{10}{13}$.

$\frac{x}{5} = \frac{10}{13}$ Original proportion

$13(x) = 5(10)$ Cross products

$13x = 50$ Simplify.

$\frac{13x}{13} = \frac{50}{13}$ Divide each side by 13.

$x = 3\frac{11}{13}$ Simplify.

Exercises

Solve each proportion. If necessary, round to the nearest hundredth.

1. $\frac{-3}{x} = \frac{2}{8}$

2. $\frac{1}{t} = \frac{5}{3}$

3. $\frac{0.1}{2} = \frac{0.5}{x}$

4. $\frac{x+1}{4} = \frac{3}{4}$

5. $\frac{4}{6} = \frac{8}{x}$

6. $\frac{x}{21} = \frac{3}{63}$

7. $\frac{9}{y+1} = \frac{18}{54}$

8. $\frac{3}{d} = \frac{18}{3}$

9. $\frac{5}{8} = \frac{p}{24}$

10. $\frac{4}{b-2} = \frac{4}{12}$

11. $\frac{1.5}{x} = \frac{12}{x}$

12. $\frac{3+y}{4} = \frac{-y}{8}$

13. $\frac{a-18}{12} = \frac{15}{3}$

14. $\frac{12}{k} = \frac{24}{k}$

15. $\frac{2+w}{6} = \frac{12}{9}$

Use a proportion to solve each problem.

16. MODELS To make a model of the Guadeloupe River bed, Hermie used 1 inch of clay for 5 miles of the river's actual length. His model river was 50 inches long. How long is the Guadeloupe River?

17. EDUCATION Josh finished 24 math problems in one hour. At that rate, how many hours will it take him to complete 72 problems?

2-7 Study Guide and Intervention

Percent of Change

Percent of Change When an increase or decrease in an amount is expressed as a percent, the percent is called the **percent of change**. If the new number is greater than the original number, the percent of change is a **percent of increase**. If the new number is less than the original number, the percent of change is the **percent of decrease**.

Example 1 Find the percent of increase.
    original: 48
    new: 60

First, subtract to find the amount of increase. The amount of increase is $60 - 48 = 12$.
Then find the percent of increase by using the original number, 48, as the base.

$$\frac{12}{48} = \frac{r}{100}$$ Percent proportion

$12(100) = 48(r)$ Cross products

$1200 = 48r$ Simplify.

$$\frac{1200}{48} = \frac{48r}{48}$$ Divide each side by 48.

$25 = r$ Simplify.

The percent of increase is 25%.

Example 2 Find the percent of decrease.
    original: 30
    new: 22

First, subtract to find the amount of decrease. The amount of decrease is $30 - 22 = 8$.
Then find the percent of decrease by using the original number, 30, as the base.

$$\frac{8}{30} = \frac{r}{100}$$ Percent proportion

$8(100) = 30(r)$ Cross products

$800 = 30r$ Simplify.

$$\frac{800}{30} = \frac{30r}{30}$$ Divide each side by 30.

$26\frac{2}{3} = r$ Simplify.

The percent of decrease is $26\frac{2}{3}\%$, or about 27%.

Exercises

State whether each percent of change is a percent of *increase* or a percent of *decrease*. Then find each percent of change. Round to the nearest whole percent.

1. original: 50
    new: 80

2. original: 90
    new: 100

3. original: 45
    new: 20

4. original: 77.5
    new: 62

5. original: 140
    new: 150

6. original: 135
    new: 90

7. original: 120
    new: 180

8. original: 90
    new: 270

9. original: 27.5
    new: 25

10. original: 84
    new: 98

11. original: 12.5
    new: 10

12. original: 250
    new: 500

2-7   Study Guide and Intervention (continued)

Percent of Change

Solve Problems Discounted prices and prices including tax are applications of percent of change. Discount is the amount by which the regular price of an item is reduced. Thus, the discounted price is an example of percent of decrease. Sales tax is an amount that is added to the cost of an item, so the price including tax is an example of percent of increase.

> **Example**   SALES A coat is on sale for 25% off the original price. If the original price of the coat is $75, what is the discounted price?
>
> The discount is 25% of the original price.
>
> 25% of $75 = 0.25×75 25% = 0.25
>
>   = 18.75   Use a calculator.
>
> Subtract $18.75 from the original price.
>
> $75 − $18.75 = $56.25
>
> The discounted price of the coat is $56.25.

Exercises

Find the total price of each item.

1. Shirt: $24.00
Sales tax: 4%

2. CD player: $142.00
Sales tax: 5.5%

3. Celebrity calendar: $10.95
Sales tax: 7.5%

Find the discounted price of each item.

4. Compact disc: $16
Discount: 15%

5. Two concert tickets: $28
Student discount: 28%

6. Airline ticket: $248.00
Superair discount: 33%

7. VIDEOS The original selling price of a new sports video was $65.00. Due to the demand the price was increased to $87.75. What was the percent of increase over the original price?

8. SCHOOL A high school paper increased its sales by 75% when it ran an issue featuring a contest to win a class party. Before the contest issue, 10% of the school's 800 students bought the paper. How many students bought the contest issue?

9. BASEBALL Baseball tickets cost $15 for general admission or $20 for box seats. The sales tax on each ticket is 8%. What is the final cost of each type of ticket?

2-8 Study Guide and Intervention

Literal Equations and Dimensional Analysis

Solve for Variables Sometimes you may want to solve an equation such as $V = \ell wh$ for one of its variables. For example, if you know the values of V, w, and h, then the equation $\ell = \dfrac{V}{wh}$ is more useful for finding the value of ℓ. If an equation that contains more than one variable is to be solved for a specific variable, use the properties of equality to isolate the specified variable on one side of the equation.

Example 1 Solve $2x - 4y = 8$, for y.

$$2x - 4y = 8$$
$$2x - 4y - 2x = 8 - 2x$$
$$-4y = 8 - 2x$$
$$\frac{-4y}{-4} = \frac{8 - 2x}{-4}$$
$$y = \frac{8 - 2x}{-4} \text{ or } \frac{2x - 8}{4}$$

The value of y is $\dfrac{2x - 8}{4}$.

Example 2 Solve $3m - n = km - 8$, for m.

$$3m - n = km - 8$$
$$3m - n - km = km - 8 - km$$
$$3m - n - km = -8$$
$$3m - n - km + n = -8 + n$$
$$3m - km = -8 + n$$
$$m(3 - k) = -8 + n$$
$$\frac{m(3 - k)}{3 - k} = \frac{-8 + n}{3 - k}$$
$$m = \frac{-8 + n}{3 - k} \text{ or } \frac{n - 8}{3 - k}$$

The value of m is $\dfrac{n - 8}{3 - k}$. Since division by 0 is undefined, $3 - k \neq 0$, or $k \neq 3$.

Exercises

Solve each equation or formula for the variable indicated.

1. $ax - b = c$, for x

2. $15x + 1 = y$, for x

3. $(x + f) + 2 = j$, for x

4. $xy + w = 9$, for y

5. $x(4 - k) = p$, for k

6. $7x + 3y = m$, for y

7. $4(r + 3) = t$, for r

8. $2x + b = w$, for x

9. $x(1 + y) = z$, for x

10. $16w + 4x = y$, for x

11. $d = rt$, for r

12. $A = \dfrac{h(a + b)}{2}$, for h

13. $C = \dfrac{5}{9}(F - 32)$, for F

14. $P = 2\ell + 2w$, for w

15. $A = \ell w$, for ℓ

2-8 Study Guide and Intervention (continued)

Literal Equations and Dimensional Analysis

Use Formulas Many real-world problems require the use of formulas. Sometimes solving a formula for a specified variable will help solve the problem.

Example The formula $C = \pi d$ represents the circumference of a circle, or the distance around the circle, where d is the diameter. If an airplane could fly around Earth at the equator without stopping, it would have traveled about 24,900 miles. Find the diameter of Earth.

$C = \pi d$ Given formula

$d = \dfrac{C}{\pi}$ Solve for d.

$d = \dfrac{24{,}900}{3.14}$ Use $\pi = 3.14$.

$d \approx 7930$ Simplify.

The diameter of Earth is about 7930 miles.

Exercises

1. **GEOMETRY** The volume of a cylinder V is given by the formula $V = \pi r^2 h$, where r is the radius and h is the height.

   a. Solve the formula for h.

   b. Find the height of a cylinder with volume 2500π cubic feet and radius 10 feet.

2. **WATER PRESSURE** The water pressure on a submerged object is given by $P = 64d$, where P is the pressure in pounds per square foot, and d is the depth of the object in feet.

   a. Solve the formula for d.

   b. Find the depth of a submerged object if the pressure is 672 pounds per square foot.

3. **GRAPHS** The equation of a line containing the points $(a, 0)$ and $(0, b)$ is given by the formula $\dfrac{x}{a} + \dfrac{y}{b} = 1$.

   a. Solve the equation for y.

   b. Suppose the line contains the points $(4, 0)$, and $(0, -2)$. If $x = 3$, find y.

4. **GEOMETRY** The surface area of a rectangular solid is given by the formula $x = 2\ell w + 2\ell h + 2wh$, where ℓ = length, w = width, and h = height.

   a. Solve the formula for h.

   b. The surface area of a rectangular solid with length 6 centimeters and width 3 centimeters is 72 square centimeters. Find the height.

2-9 Study Guide and Intervention

Weighted Averages

Mixture Problems **Mixture Problems** are problems where two or more parts are combined into a whole. They involve weighted averages. In a mixture problem, the weight is usually a price or a percent of something.

Weighted Average	The weighted average M of a set of data is the sum of the product of each number in the set and its weight divided by the sum of all the weights.

Example **COOKIES** Delectable Cookie Company sells chocolate chip cookies for $6.95 per pound and white chocolate cookies for $5.95 per pound. How many pounds of chocolate chip cookies should be mixed with 4 pounds of white chocolate cookies to obtain a mixture that sells for $6.75 per pound.

Let w = the number of pounds of chocolate chip cookies

	Number of Pounds	Price per Pound	Total Price
Chocolate Chip	w	6.95	$6.95w$
White Chocolate	4	5.95	$4(5.95)$
Mixture	$w + 4$	6.75	$6.75(w + 4)$

Equation: $6.95w + 4(5.95) = 6.75(w + 4)$

Solve the equation.

$6.95w + 4(5.95) = 6.75(w + 4)$	Original equation
$6.95w + 23.80 = 6.75w + 27$	Simplify.
$6.95w + 23.80 - 6.75w = 6.75w + 27 - 6.75w$	Subtract 6.75w from each side.
$0.2w + 23.80 = 27$	Simplify.
$0.2w + 23.80 - 23.80 = 27 - 23.80$	Subtract 23.80 from each side.
$0.2w = 3.2$	Simplify.
$w = 16$	Simplify.

16 pounds of chocolate chip cookies should be mixed with 4 pounds of white chocolate cookies.

Exercises

1. **SOLUTIONS** How many grams of sugar must be added to 60 grams of a solution that is 32% sugar to obtain a solution that is 50% sugar?

2. **NUTS** The Quik Mart has two kinds of nuts. Pecans sell for $1.55 per pound and walnuts sell for $1.95 per pound. How many pounds of walnuts must be added to 15 pounds of pecans to make a mixture that sells for $1.75 per pound?

3. **INVESTMENTS** Alice Gleason invested a portion of $32,000 at 9% interest and the balance at 11% interest. How much did she invest at each rate if her total income from both investments was $3200.

4. **MILK** Whole milk is 4% butterfat. How much skim milk with 0% butterfat should be added to 32 ounces of whole milk to obtain a mixture that is 2.5% butterfat?

2-9 **Study Guide and Intervention** (continued)

Weighted Averages

Uniform Motion Problems Motion problems are another application of weighted averages. **Uniform motion problems** are problems where an object moves at a certain speed, or rate. Use the formula $d = rt$ to solve these problems, where d is the distance, r is the rate, and t is the time.

> **Example**   **DRIVING** Bill Gutierrez drove at a speed of 65 miles per hour on an expressway for 2 hours. He then drove for 1.5 hours at a speed of 45 miles per hour on a state highway. What was his average speed?
>
> $M = \dfrac{65 \cdot 2 + 45 \cdot 1.5}{2 + 1.5}$ Definition of weighted average
>
> ≈ 56.4 Simplify.
>
> Bill drove at an average speed of about 56.4 miles per hour.

Exercises

1. **TRAVEL** Mr. Anders and Ms. Rich each drove home from a business meeting. Mr. Anders traveled east at 100 kilometers per hour and Ms. Rich traveled west at 80 kilometers per hours. In how many hours were they 100 kilometers apart.

2. **AIRPLANES** An airplane flies 750 miles due west in $1\frac{1}{2}$ hours and 750 miles due south in 2 hours. What is the average speed of the airplane?

3. **TRACK** Sprinter A runs 100 meters in 15 seconds, while sprinter B starts 1.5 seconds later and runs 100 meters in 14 seconds. If each of them runs at a constant rate, who is further in 10 seconds after the start of the race? Explain.

4. **TRAINS** An express train travels 90 kilometers per hour from Smallville to Megatown. A local train takes 2.5 hours longer to travel the same distance at 50 kilometers per hour. How far apart are Smallville and Megatown?

5. **CYCLING** Two cyclists begin traveling in the same direction on the same bike path. One travels at 15 miles per hour, and the other travels at 12 miles per hour. When will the cyclists be 10 miles apart?

6. **TRAINS** Two trains leave Chicago, one traveling east at 30 miles per hour and one traveling west at 40 miles per hour. When will the trains be 210 miles apart?

3-1 Study Guide and Intervention

Graphing Linear Equations

Identify Linear Equations and Intercepts

A **linear equation** is an equation that can be written in the form $Ax + By = C$. This is called the **standard form** of a linear equation.

Standard Form of a Linear Equation	$Ax + By = C$, where $A \geq 0$, A and B are not both zero, and A, B, and C are integers with a greatest common factor of 1

Example 1 Determine whether $y = 6 - 3x$ is a linear equation. Write the equation in standard form.

First rewrite the equation so both variables are on the same side of the equation.

$y = 6 - 3x$	Original equation
$y + 3x = 6 - 3x + 3x$	Add 3x to each side.
$3x + y = 6$	Simplify.

The equation is now in standard form, with $A = 3$, $B = 1$ and $C = 6$. This is a linear equation.

Example 2 Determine whether $3xy + y = 4 + 2x$ is a linear equation. Write the equation in standard form.

Since the term $3xy$ has two variables, the equation cannot be written in the form $Ax + By = C$. Therefore, this is not a linear equation.

Exercises

Determine whether each equation is a linear equation. Write *yes* or *no*. If yes, write the equation in standard form.

1. $2x = 4y$

2. $6 + y = 8$

3. $4x - 2y = -1$

4. $3xy + 8 = 4y$

5. $3x - 4 = 12$

6. $y = x^2 + 7$

7. $y - 4x = 9$

8. $x + 8 = 0$

9. $-2x + 3 = 4y$

10. $2 + \frac{1}{2}x = y$

11. $\frac{1}{4}y = 12 - 4x$

12. $3xy - y = 8$

13. $6x + 4y - 3 = 0$

14. $yx - 2 = 8$

15. $6x - 2y = 8 + y$

16. $\frac{1}{4}x - 12y = 1$

17. $3 + x + x^2 = 0$

18. $x^2 = 2xy$

3-1 Study Guide and Intervention (continued)

Graphing Linear Equations

Graph Linear Equations The graph of a linear equations represents all the solutions of the equation. An x-coordinate of the point at which a graph of an equation crosses the x-axis in an **x-intercept**. A y-coordinate of the point at which a graph crosses the y-axis is called a **y-intercept**.

Example 1 Graph $3x + 2y = 6$ by using the x- and y-intercepts.

To find the x-intercept, let $y = 0$ and solve for x. The x-intercept is 2. The graph intersects the x-axis at $(2, 0)$.

To find the y-intercept, let $x = 0$ and solve for y.

The y-intercept is 3. The graph intersects the y-axis at $(0, 3)$.

Plot the points $(2, 0)$ and $(0, 3)$ and draw the line through them.

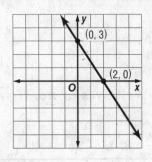

Example 2 Graph $y - 2x = 1$ by making a table.

Solve the equation for y.

$$y - 2x = 1 \qquad \text{Original equation}$$
$$y - 2x + 2x = 1 + 2x \qquad \text{Add } 2x \text{ to each side.}$$
$$y = 2x + 1 \qquad \text{Simplify.}$$

Select five values for the domain and make a table. Then graph the ordered pairs and draw a line through the points.

x	$2x + 1$	y	(x, y)
-2	$2(-2) + 1$	-3	$(-2, -3)$
-1	$2(-1) + 1$	-1	$(-1, -1)$
0	$2(0) + 1$	1	$(0, 1)$
1	$2(1) + 1$	3	$(1, 3)$
2	$2(2) + 1$	5	$(2, 5)$

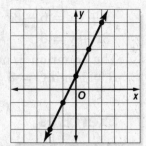

Exercises

Graph each equation by using the x- and y-intercepts.

1. $2x + y = -2$

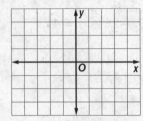

2. $3x - 6y = -3$

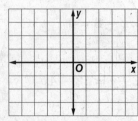

3. $-2x + y = -2$

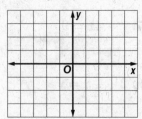

Graph each equation by making a table.

4. $y = 2x$

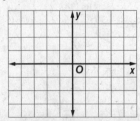

5. $x - y = -1$

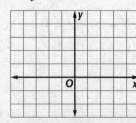

6. $x + 2y = 4$

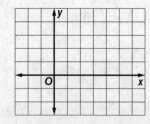

3-2 Study Guide and Intervention

Solving Linear Equations by Graphing

Solve by Graphing You can solve an equation by graphing the related function. The solution of the equation is the x-intercept of the function.

Example **Solve the equation $2x - 2 = -4$ by graphing.**

First set the equation equal to 0. Then replace 0 with $f(x)$. Make a table of ordered pair solutions. Graph the function and locate the x-intercept.

$2x - 2 = -4$	Original equation
$2x - 2 + 4 = -4 + 4$	Add 4 to each side.
$2x + 2 = 0$	Simplify.
$f(x) = 2x + 2$	Replace 0 with $f(x)$.

To graph the function, make a table. Graph the ordered pairs.

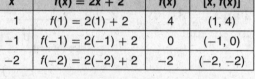

x	$f(x) = 2x + 2$	$f(x)$	$[x, f(x)]$
1	$f(1) = 2(1) + 2$	4	$(1, 4)$
-1	$f(-1) = 2(-1) + 2$	0	$(-1, 0)$
-2	$f(-2) = 2(-2) + 2$	-2	$(-2, -2)$

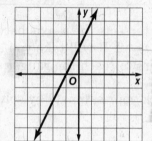

The graph intersects the x-axis at $(-1, 0)$.

The solution to the equation is $x = -1$.

Exercises

Solve each equation.

1. $3x - 3 = 0$

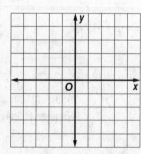

2. $-2x + 1 = 5 - 2x$

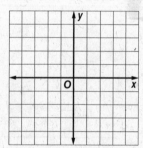

3. $-x + 4 = 0$

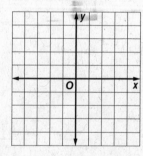

4. $0 = 4x - 1$

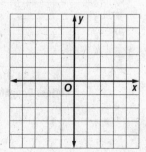

5. $5x - 1 = 5x$

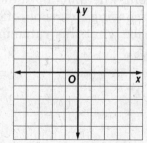

6. $-3x + 1 = 0$

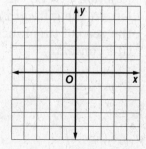

3-2 Study Guide and Intervention (continued)

Solving Linear Equations by Graphing

Estimate Solutions by Graphing Sometimes graphing does not provide an exact solution, but only an estimate. In these cases, solve algebraically to find the exact solution.

Example **WALKING** You and your cousin decide to walk the 7-mile trail at the state park to the ranger station. The function $d = 7 - 3.2t$ represents your distance d from the ranger station after t hours. Find the zero of this function. Describe what this value means in this context.

Make a table of values to graph the function.

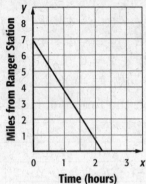

t	$d = 7 - 3.2t$	d	(t, d)
0	$d = 7 - 3.2(0)$	7	$(0, 7)$
1	$d = 7 - 3.2(1)$	3.8	$(1, 3.8)$
2	$d = 7 - 3.2(2)$	0.6	$(2, 0.6)$

The graph intersects the t–axis between $t = 2$ and $t = 3$, but closer to $t = 2$. It will take you and your cousin just over two hours to reach the ranger station.

You can check your estimate by solving the equation algebraically.

Exercises

1. **MUSIC** Jessica wants to record her favorite songs to one CD. The function $C = 80 - 3.22n$ represents the recording time C available after n songs are recorded. Find the zero of this function. Describe what this value means in this context.

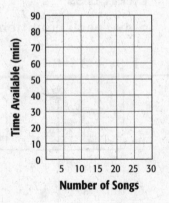

2. **GIFT CARDS** Enrique uses a gift card to buy coffee at a coffee shop. The initial value of the gift card is $20. The function $n = 20 - 2.75c$ represents the amount of money still left on the gift card n after purchasing c cups of coffee. Find the zero of this function. Describe what this value means in this context.

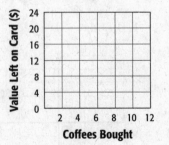

3-3 Study Guide and Intervention

Rate of Change and Slope

Rate of Change The **rate of change** tells, on average, how a quantity is changing over time.

> **Example** **POPULATION The graph shows the population growth in China.**
>
> **a. Find the rates of change for 1950–1975 and for 2000–2025.**
>
> 1950–1975: $\dfrac{\text{change in population}}{\text{change in time}} = \dfrac{0.93 - 0.55}{1975 - 1950}$
>
> $\qquad\qquad = \dfrac{0.38}{25}$ or 0.0152
>
> 2000–2025: $\dfrac{\text{change in population}}{\text{change in time}} = \dfrac{1.45 - 1.27}{2025 - 2000}$
>
> $\qquad\qquad = \dfrac{0.18}{25}$ or 0.0072

Population Growth in China

People (billions) vs Year

0.55 (1950), 0.93 (1975), 1.27 (2000), 1.45 (2025*)

*Estimated

Source: United Nations Population Division

b. Explain the meaning of the rate of change in each case.

From 1950–1975, the growth was 0.0152 billion per year, or 15.2 million per year. From 2000–2025, the growth is expected to be 0.0072 billion per year, or 7.2 million per year.

c. How are the different rates of change shown on the graph?

There is a greater vertical change for 1950–1975 than for 2000–2025. Therefore, the section of the graph for 1950–1975 has a steeper slope.

Exercises

1. LONGEVITY The graph shows the predicted life expectancy for men and women born in a given year.

a. Find the rates of change for women from 2000–2025 and 2025–2050.

b. Find the rates of change for men from 2000–2025 and 2025–2050.

c. Explain the meaning of your results in Exercises 1 and 2.

d. What pattern do you see in the increase with each 25-year period?

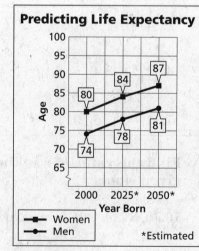

Predicting Life Expectancy

Age vs Year Born

Women: 80 (2000), 84 (2025*), 87 (2050*)
Men: 74 (2000), 78 (2025*), 81 (2050*)

*Estimated

Source: USA TODAY

e. Make a prediction for the life expectancy for 2050–2075. Explain how you arrived at your prediction.

3-3 Study Guide and Intervention (continued)

Rate of Change and Slope

Find Slope The **slope** of a line is the ratio of change in the y-coordinates (rise) to the change in the x-coordinates (run) as you move in the positive direction.

Slope of a Line	$m = \dfrac{\text{rise}}{\text{run}}$ or $m = \dfrac{y_2 - y_1}{x_2 - x_1}$, where (x_1, y_1) and (x_2, y_2) are the coordinates of any two points on a nonvertical line

---

Example 1 Find the slope of the line that passes through $(-3, 5)$ and $(4, -2)$.

Let $(-3, 5) = (x_1, y_1)$ and $(4, -2) = (x_2, y_2)$.

$m = \dfrac{y_2 - y_1}{x_2 - x_1}$ Slope formula

$= \dfrac{-2 - 5}{4 - (-3)}$ $y_2 = -2, y_1 = 5, x_2 = 4, x_1 = -3$

$= \dfrac{-7}{7}$ Simplify.

$= -1$

Example 2 Find the value of r so that the line through $(10, r)$ and $(3, 4)$ has a slope of $-\dfrac{2}{7}$.

$m = \dfrac{y_2 - y_1}{x_2 - x_1}$ Slope formula

$-\dfrac{2}{7} = \dfrac{4 - r}{3 - 10}$ $m = -\dfrac{2}{7}, y_2 = 4, y_1 = r, x_2 = 3, x_1 = 10$

$-\dfrac{2}{7} = \dfrac{4 - r}{-7}$ Simplify.

$-2(-7) = 7(4 - r)$ Cross multiply.

$14 = 28 - 7r$ Distributive Property

$-14 = -7r$ Subtract 28 from each side.

$2 = r$ Divide each side by −7.

Exercises

Find the slope of the line that passes through each pair of points.

1. $(4, 9), (1, 6)$

2. $(-4, -1), (-2, -5)$

3. $(-4, -1), (-4, -5)$

4. $(2, 1), (8, 9)$

5. $(14, -8), (7, -6)$

6. $(4, -3), (8, -3)$

7. $(1, -2), (6, 2)$

8. $(2, 5), (6, 2)$

9. $(4, 3.5), (-4, 3.5)$

Find the value of r so the line that passes through each pair of points has the given slope.

10. $(6, 8), (r, -2), m = 1$

11. $(-1, -3), (7, r), m = \dfrac{3}{4}$

12. $(2, 8), (r, -4)\ m = -3$

13. $(7, -5), (6, r), m = 0$

14. $(r, 4), (7, 1), m = \dfrac{3}{4}$

15. $(7, 5), (r, 9), m = 6$

3-4 Study Guide and Intervention

Direct Variation

Direct Variation Equations A **direct variation** is described by an equation of the form $y = kx$, where $k \neq 0$. We say that y *varies directly as x*. In the equation $y = kx$, k is the **constant of variation**.

Example 1 Name the constant of variation for the equation. Then find the slope of the line that passes through the pair of points.

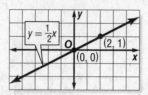

For $y = \frac{1}{2}x$, the constant of variation is $\frac{1}{2}$.

$$m = \frac{y_2 - y_1}{x_2 - x_1} \quad \text{Slope formula}$$

$$= \frac{1 - 0}{2 - 0} \quad (x_1, y_1) = (0, 0), (x_2, y_2) = (2, 1)$$

$$= \frac{1}{2} \quad \text{Simplify.}$$

The slope is $\frac{1}{2}$.

Example 2 Suppose y varies directly as x, and $y = 30$ when $x = 5$.

a. Write a direct variation equation that relates x and y.

Find the value of k.

$y = kx$ Direct variation equation
$30 = k(5)$ Replace y with 30 and x with 5.
$6 = k$ Divide each side by 5.

Therefore, the equation is $y = 6x$.

b. Use the direct variation equation to find x when $y = 18$.

$y = 6x$ Direct variation equation
$18 = 6x$ Replace y with 18.
$3 = x$ Divide each side by 6.

Therefore, $x = 3$ when $y = 18$.

Exercises

Name the constant of variation for each equation. Then determine the slope of the line that passes through each pair of points.

1.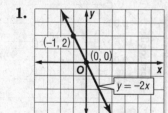

2.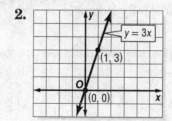

3.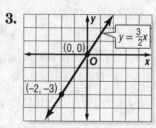

Suppose y varies directly as x. Write a direct variation equation that relates x to y. Then solve.

4. If $y = 4$ when $x = 2$, find y when $x = 16$.

5. If $y = 9$ when $x = -3$, find x when $y = 6$.

6. If $y = -4.8$ when $x = -1.6$, find x when $y = -24$.

7. If $y = \frac{1}{4}$ when $x = \frac{1}{8}$, find x when $y = \frac{3}{16}$.

3-4 Study Guide and Intervention (continued)

Direct Variation

Direct Variation Problems The **distance formula** $d = rt$ is a direct variation equation. In the formula, distance d varies directly as time t, and the rate r is the constant of variation.

Example | **TRAVEL A family drove their car 225 miles in 5 hours.**

a. Write a direct variation equation to find the distance traveled for any number of hours.

Use given values for d and t to find r.

$d = rt$	Original equation
$225 = r(5)$	$d = 225$ and $t = 5$
$45 = r$	Divide each side by 5.

Therefore, the direct variation equation is $d = 45t$.

b. Graph the equation.

The graph of $d = 45t$ passes through the origin with slope 45.

$$m = \frac{45}{1} \qquad \frac{\text{rise}}{\text{run}}$$

✓**CHECK** (5, 225) lies on the graph.

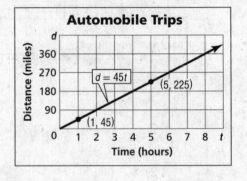

Automobile Trips

c. Estimate how many hours it would take the family to drive 360 miles.

$d = 45t$	Original equation
$360 = 45t$	Replace d with 360.
$t = 8$	Divide each side by 45.

Therefore, it will take 8 hours to drive 360 miles.

Exercises

1. **RETAIL** The total cost C of bulk jelly beans is $4.49 times the number of pounds p.

   a. Write a direct variation equation that relates the variables.

   b. Graph the equation on the grid at the right.

   c. Find the cost of $\frac{3}{4}$ pound of jelly beans.

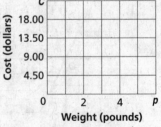

Cost of Jelly Beans

2. **CHEMISTRY** Charles's Law states that, at a constant pressure, volume of a gas V varies directly as its temperature T. A volume of 4 cubic feet of a certain gas has a temperature of 200 degrees Kelvin.

   a. Write a direct variation equation that relates the variables.

   b. Graph the equation on the grid at the right.

   c. Find the volume of the same gas at 250 degrees Kelvin.

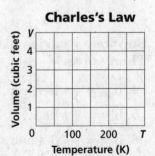

Charles's Law

3-5 Study Guide and Intervention

Arithmetic Sequences as Linear Functions

Recognize Arithmetic Sequences A **sequence** is a set of numbers in a specific order. If the difference between successive terms is constant, then the sequence is called an **arithmetic sequence**.

Arithmetic Sequence	a numerical pattern that increases or decreases at a constant rate or value called the **common difference**
Terms of an Arithmetic Sequence	If a_1 is the first term of an arithmetic sequence with common difference d, then the sequence is $a_1, a_1 + d, a_1 + 2d, a_1 + 3d, \ldots$.
nth Term of an Arithmetic Sequence	$a_n = a_1 + (n - 1)d$

Example 1 Determine whether the sequence 1, 3, 5, 7, 9, 11, ... is an arithmetic sequence. Justify your answer.

If possible, find the common difference between the terms. Since $3 - 1 = 2$, $5 - 3 = 2$, and so on, the common difference is 2.

Since the difference between the terms of 1, 3, 5, 7, 9, 11, ... is constant, this is an arithmetic sequence.

Example 2 Write an equation for the nth term of the sequence 12, 15, 18, 21, ...

In this sequence, a_1 is 12. Find the common difference.

12   15   18   21
  +3   +3   +3

The common difference is 3.

Use the formula for the nth term to write an equation.

$a_n = a_1 + (n - 1)d$ Formula for the nth term

$a_n = 12 + (n - 1)3$ $a_1 = 12, d = 3$

$a_n = 12 + 3n - 3$ Distributive Property

$a_n = 3n + 9$ Simplify.

The equation for the nth term is $a_n = 3n + 9$.

Exercises

Determine whether each sequence is an arithmetic sequence. Write *yes* or *no*. Explain.

1. 1, 5, 9, 13, 17, ...

2. 8, 4, 0, −4, −8, ...

3. 1, 3, 9, 27, 81, ...

Find the next three terms of each arithmetic sequence.

4. 9, 13, 17, 21, 25, ...

5. 4, 0, −4, −8, −12, ...

6. 29, 35, 41, 47, ...

Write an equation for the nth term of each arithmetic sequence. Then graph the first five terms of the sequence.

7. 1, 3, 5, 7, ...

8. −1, −4, −7, −10, ...

9. −4, −9, −14, −19, ...

3-5 Study Guide and Intervention (continued)

Arithmetic Sequences as Linear Functions

Arithmetic Sequences and Functions An **arithmetic sequence** is a linear function in which n is the independent variable, a_n is the dependent variable, and the common difference d is the slope. The formula can be rewritten as the function $a_n = a_1 + (n - 1)d$, where n is a counting number.

> **Example** **SEATING** There are 20 seats in the first row of the balcony of the auditorium. There are 22 seats in the second row, and 24 seats in the third row.

a. Write a function to represent this sequence.

The first term a_1 is 20. Find the common difference.

$$20 \quad 22 \quad 24$$
$$\ \ +2 \quad +2$$

The common difference is 2.

$a_n = a_1 + (n - 1)d$ Formula for the nth term

$ = 20 + (n - 1)2$ $a_1 = 20$ and $d = 2$

$ = 20 + 2n - 2$ Distributive Property

$ = 18 + 2n$ Simplify.

The function is $a_n = 18 + 2n$.

b. Graph the function.

The rate of change is 2. Make a table and plot points.

n	a_n
1	20
2	22
3	24
4	26

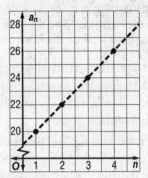

Exercises

1. **KNITTING** Sarah learns to knit from her grandmother. Two days ago, she measured the length of the scarf she is knitting to be 13 inches. Yesterday, she measured the length of the scarf to be 15.5 inches. Today it measures 18 inches. Write a function to represent the arithmetic sequence.

2. **REFRESHMENTS** You agree to pour water into the cups for the Booster Club at a football game. The pitcher contains 64 ounces of water when you begin. After you have filled 8 cups, the pitcher is empty and must be refilled.

   a. Write a function to represent the arithmetic sequence.

   b. Graph the function.

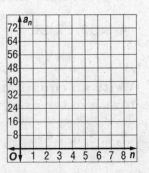

     44

3-6 Study Guide and Intervention

Proportional and Nonproportional Relationships

Proportional Relationships If the relationship between the domain and range of a relation is linear, the relationship can be described by a linear equation. If the equation passes through $(0, 0)$ and is of the form $y = kx$, then the relationship is proportional.

Example **COMPACT DISCS** Suppose you purchased a number of packages of blank compact discs. If each package contains 3 compact discs, you could make a chart to show the relationship between the number of packages of compact discs and the number of discs purchased. Use x for the number of packages and y for the number of compact discs.

Make a table of ordered pairs for several points of the graph.

Number of Packages	1	2	3	4	5
Number of CDs	3	6	9	12	15

The difference in the x values is 1, and the difference in the y values is 3. This pattern shows that y is always three times x. This suggests the relation $y = 3x$. Since the relation is also a function, we can write the equation in function notation as $f(x) = 3x$.

The relation includes the point $(0, 0)$ because if you buy 0 packages of compact discs, you will not have any compact discs. Therefore, the relationship is proportional.

Exercises

1. **NATURAL GAS** Natural gas use is often measured in "therms." The total amount a gas company will charge for natural gas use is based on how much natural gas a household uses. The table shows the relationship between natural gas use and the total cost.

Gas Used (therms)	1	2	3	4
Total Cost ($)	$1.30	$2.60	$3.90	$5.20

a. Graph the data. What can you deduce from the pattern about the relationship between the number of therms used and the total cost?

b. Write an equation to describe this relationship.

c. Use this equation to predict how much it will cost if a household uses 40 therms.

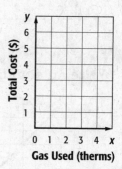

3-6 Study Guide and Intervention (continued)

Proportional and Nonproportional Relationships

Nonproportional Relationships If the ratio of the value of x to the value of y is different for select ordered pairs on the line, the equation is nonproportional.

Example Write an equation in functional notation for the relation shown in the graph.

Select points from the graph and place them in a table.

x	−1	0	1	2	3
y	4	2	0	−2	−4

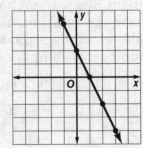

The difference between the x–values is 1, while the difference between the y-values is −2. This suggests that $y = -2x$.

If $x = 1$, then $y = -2(1)$ or −2. But the y–value for $x = 1$ is 0.

x	1	2	3
−2x	−2	−4	−6
y	0	−2	−4

y is always 2 more than −2x

This pattern shows that 2 should be added to one side of the equation. Thus, the equation is $y = -2x + 2$.

Exercises

Write an equation in function notation for the relation shown in the table. Then complete the table.

1.
x	−1	0	1	2	3	4
y	−2	2	6			

2.
x	−2	−1	0	1	2	3
y	10	7	4			

Write an equation in function notation for each relation.

3.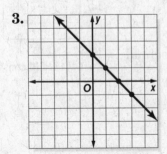

4.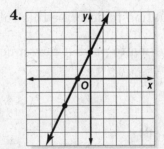

4-1 Study Guide and Intervention

Graphing Equations in Slope-Intercept Form

Slope-Intercept Form

Slope-Intercept Form	$y = mx + b$, where m is the slope and b is the y-intercept

Example 1 Write an equation in slope-intercept form for the line with a slope of -4 and a y-intercept of 3.

$y = mx + b$ Slope-intercept form

$y = -4x + 3$ Replace m with -4 and b with 3.

Example 2 Graph $3x - 4y = 8$.

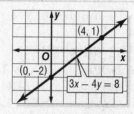

$3x - 4y = 8$ Original equation

$-4y = -3x + 8$ Subtract $3x$ from each side.

$\dfrac{-4y}{-4} = \dfrac{-3x + 8}{-4}$ Divide each side by -4.

$y = \dfrac{3}{4}x - 2$ Simplify.

The y-intercept of $y = \dfrac{3}{4}x - 2$ is -2 and the slope is $\dfrac{3}{4}$. So graph the point $(0, -2)$. From this point, move up 3 units and right 4 units. Draw a line passing through both points.

Exercises

Write an equation of a line in slope-intercept form with the given slope and y-intercept.

1. slope: 8, y-intercept -3 **2.** slope: -2, y-intercept -1 **3.** slope: -1, y-intercept -7

Write an equation in slope-intercept form for each graph shown.

4.

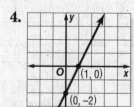

5.

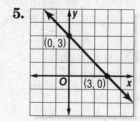

6.

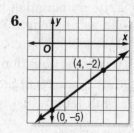

Graph each equation.

7. $y = 2x + 1$ **8.** $y = -3x + 2$ **9.** $y = -x - 1$

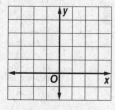   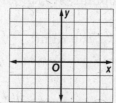   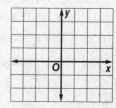

4-1 Study Guide and Intervention (continued)

Graphing Equations in Slope-Intercept Form

Modeling Real-World Data

Example MEDIA Since 1999, the number of music cassettes sold has decreased by an average rate of 27 million per year. There were 124 million music cassettes sold in 1999.

a. Write a linear equation to find the average number of music cassettes sold in any year after 1999.

The rate of change is -27 million per year. In the first year, the number of music cassettes sold was 124 million. Let $N =$ the number of millions of music cassettes sold. Let $x =$ the number of years since 1999. An equation is $N = -27x + 124$.

b. Graph the equation.

The graph of $N = -27x + 124$ is a line that passes through the point at $(0, 124)$ and has a slope of -27.

c. Find the approximate number of music cassettes sold in 2003.

$N = -27x + 124$	Original equation
$N = -27(4) + 124$	Replace x with 4.
$N = 16$	Simplify.

There were about 16 million music cassettes sold in 2003.

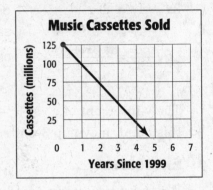

Music Cassettes Sold

Exercises

1. **MUSIC** In 2001, full-length cassettes represented 3.4% of total music sales. Between 2001 and 2006, the percent decreased by about 0.5% per year.

   a. Write an equation to find the percent P of recorded music sold as full-length cassettes for any year x between 2001 and 2006.

   b. Graph the equation on the grid at the right.

   c. Find the percent of recorded music sold as full-length cassettes in 2004.

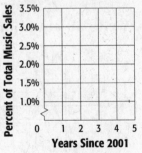

Full-length Cassette Sales

2. **POPULATION** The population of the United States is projected to be 300 million by the year 2010. Between 2010 and 2050, the population is expected to increase by about 2.5 million per year.

   a. Write an equation to find the population P in any year x between 2010 and 2050.

   b. Graph the equation on the grid at the right.

   c. Find the population in 2050.

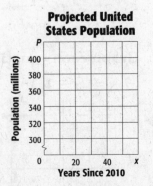

Projected United States Population

4-2 Study Guide and Intervention

Writing Equations in Slope-Intercept Form

Write an Equation Given the Slope and a Point

Example 1 Write an equation of the line that passes through $(-4, 2)$ with a slope of 3.

The line has slope 3. To find the y-intercept, replace m with 3 and (x, y) with $(-4, 2)$ in the slope-intercept form. Then solve for b.

$y = mx + b$ Slope-intercept form
$2 = 3(-4) + b$ $m = 3, y = 2$, and $x = -4$
$2 = -12 + b$ Multiply.
$14 = b$ Add 12 to each side.

Therefore, the equation is $y = 3x + 14$.

Example 2 Write an equation of the line that passes through $(-2, -1)$ with a slope of $\frac{1}{4}$.

The line has slope $\frac{1}{4}$. Replace m with $\frac{1}{4}$ and (x, y) with $(-2, -1)$ in the slope-intercept form.

$y = mx + b$ Slope-intercept form
$-1 = \frac{1}{4}(-2) + b$ $m = \frac{1}{4}, y = -1$, and $x = -2$
$-1 = -\frac{1}{2} + b$ Multiply.
$-\frac{1}{2} = b$ Add $\frac{1}{2}$ to each side.

Therefore, the equation is $y = \frac{1}{4}x - \frac{1}{2}$.

Exercises

Write an equation of the line that passes through the given point and has the given slope.

1.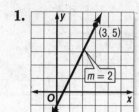

2.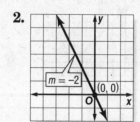

3.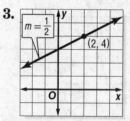

4. $(8, 2)$; slope $-\frac{3}{4}$

5. $(-1, -3)$; slope 5

6. $(4, -5)$; slope $-\frac{1}{2}$

7. $(-5, 4)$; slope 0

8. $(2, 2)$; slope $\frac{1}{2}$

9. $(1, -4)$; slope -6

10. $(-3, 0)$, $m = 2$

11. $(0, 4)$, $m = -3$

12. $(0, 350)$, $m = \frac{1}{5}$

4-2   Study Guide and Intervention *(continued)*

Writing Equations in Slope-Intercept Form

Write an Equation Given Two Points

Example   **Write an equation of the line that passes through (1, 2) and (3, −2).**
Find the slope m. To find the y-intercept, replace m with its computed value and (x, y) with (1, 2) in the slope-intercept form. Then solve for b.

$m = \dfrac{y_2 - y_1}{x_2 - x_1}$ Slope formula

$m = \dfrac{-2 - 2}{3 - 1}$ $y_2 = -2, y_1 = 2, x_2 = 3, x_1 = 1$

$m = -2$ Simplify.

$y = mx + b$ Slope-intercept form

$2 = -2(1) + b$ Replace m with −2, y with 2, and x with 1.

$2 = -2 + b$ Multiply.

$4 = b$ Add 2 to each side.

Therefore, the equation is $y = -2x + 4$.

Exercises

Write an equation of the line that passes through each pair of points.

1.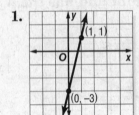
    2.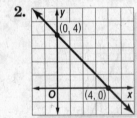
    3.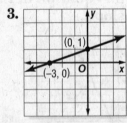

4. (−1, 6), (7, −10)   **5.** (0, 2), (1, 7)   **6.** (6, −25), (−1, 3)

7. (−2, −1), (2, 11)   **8.** (10, −1), (4, 2)   **9.** (−14, −2), (7, 7)

10. (4, 0), (0, 2)   **11.** (−3, 0), (0, 5)   **12.** (0, 16), (−10, 0)

4-3 Study Guide and Intervention

Writing Equations in Point-Slope Form

Point-Slope Form

Point-Slope Form	$y - y_1 = m(x - x_1)$, where (x_1, y_1) is a given point on a nonvertical line and m is the slope of the line

Example 1 Write an equation in point-slope form for the line that passes through (6, 1) with a slope of $-\frac{5}{2}$.

$y - y_1 = m(x - x_1)$ Point-slope form

$y - 1 = -\frac{5}{2}(x - 6)$ $m = -\frac{5}{2}$; $(x_1, y_1) = (6, 1)$

Therefore, the equation is $y - 1 = -\frac{5}{2}(x - 6)$.

Example 2 Write an equation in point-slope form for a horizontal line that passes through (4, −1).

$y - y_1 = m(x - x_1)$ Point-slope form

$y - (-1) = 0(x - 4)$ $m = 0$; $(x_1, y_1) = (4, -1)$

$y + 1 = 0$ Simplify.

Therefore, the equation is $y + 1 = 0$.

Exercises

Write an equation in point-slope form for the line that passes through each point with the given slope.

1.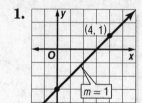

2.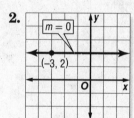

3.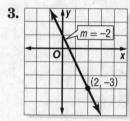

4. $(2, 1)$, $m = 4$

5. $(-7, 2)$, $m = 6$

6. $(8, 3)$, $m = 1$

7. $(-6, 7)$, $m = 0$

8. $(4, 9)$, $m = \frac{3}{4}$

9. $(-4, -5)$, $m = -\frac{1}{2}$

10. Write an equation in point-slope form for a horizontal line that passes through (4, −2).

11. Write an equation in point-slope form for a horizontal line that passes through (−5, 6).

12. Write an equation in point-slope form for a horizontal line that passes through (5, 0).

4-3 Study Guide and Intervention *(continued)*

Writing Equations in Point-Slope Form

Forms of Linear Equations

Slope-Intercept Form	$y = mx + b$	m = slope; b = y-intercept
Point-Slope Form	$y - y_1 = m(x - x_1)$	m = slope; (x_1, y_1) is a given point
Standard Form	$Ax + By = C$	A and B are not both zero. Usually A is nonnegative and A, B, and C are integers whose greatest common factor is 1.

Example 1 Write $y + 5 = \frac{2}{3}(x - 6)$ in standard form.

$y + 5 = \frac{2}{3}(x - 6)$ Original equation

$3(y + 5) = 3\left(\frac{2}{3}\right)(x - 6)$ Multiply each side by 3.

$3y + 15 = 2(x - 6)$ Distributive Property

$3y + 15 = 2x - 12$ Distributive Property

$3y = 2x - 27$ Subtract 15 from each side.

$-2x + 3y = -27$ Add $-2x$ to each side.

$2x - 3y = 27$ Multiply each side by -1.

Therefore, the standard form of the equation is $2x - 3y = 27$.

Example 2 Write $y - 2 = -\frac{1}{4}(x - 8)$ in slope-intercept form.

$y - 2 = -\frac{1}{4}(x - 8)$ Original equation

$y - 2 = -\frac{1}{4}x + 2$ Distributive Property

$y = -\frac{1}{4}x + 4$ Add 2 to each side.

Therefore, the slope-intercept form of the equation is $y = -\frac{1}{4}x + 4$.

Exercises

Write each equation in standard form.

1. $y + 2 = -3(x - 1)$ **2.** $y - 1 = -\frac{1}{3}(x - 6)$ **3.** $y + 2 = \frac{2}{3}(x - 9)$

4. $y + 3 = -(x - 5)$ **5.** $y - 4 = \frac{5}{3}(x + 3)$ **6.** $y + 4 = -\frac{2}{5}(x - 1)$

Write each equation in slope-intercept form.

7. $y + 4 = 4(x - 2)$ **8.** $y - 5 = \frac{1}{3}(x - 6)$ **9.** $y - 8 = -\frac{1}{4}(x + 8)$

10. $y - 6 = 3\left(x - \frac{1}{3}\right)$ **11.** $y + 4 = -2(x + 5)$ **12.** $y + \frac{5}{3} = \frac{1}{2}(x - 2)$

4-4 Study Guide and Intervention

Parallel and Perpendicular Lines

Parallel Lines Two nonvertical lines are **parallel** if they have the same slope. All vertical lines are parallel.

Example Write an equation in slope-intercept form for the line that passes through $(-1, 6)$ and is parallel to the graph of $y = 2x + 12$.

A line parallel to $y = 2x + 12$ has the same slope, 2. Replace m with 2 and (x_1, y_1) with $(-1, 6)$ in the point-slope form.

$y - y_1 = m(x - x_1)$ Point-slope form

$y - 6 = 2(x - (-1))$ $m = 2; (x_1, y_1) = (-1, 6)$

$y - 6 = 2(x + 1)$ Simplify.

$y - 6 = 2x + 2$ Distributive Property

$y = 2x + 8$ Slope-intercept form

Therefore, the equation is $y = 2x + 8$.

Exercises

Write an equation in slope-intercept form for the line that passes through the given point and is parallel to the graph of each equation.

1.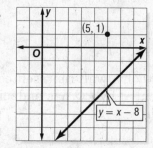

2.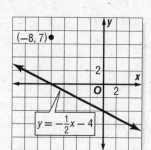

3.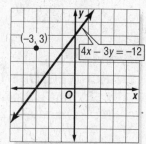

4. $(-2, 2), y = 4x - 2$

5. $(6, 4), y = \frac{1}{3}x + 1$

6. $(4, -2), y = -2x + 3$

7. $(-2, 4), y = -3x + 10$

8. $(-1, 6), 3x + y = 12$

9. $(4, -6), x + 2y = 5$

10. Find an equation of the line that has a y-intercept of 2 that is parallel to the graph of the line $4x + 2y = 8$.

11. Find an equation of the line that has a y-intercept of -1 that is parallel to the graph of the line $x - 3y = 6$.

12. Find an equation of the line that has a y-intercept of -4 that is parallel to the graph of the line $y = 6$.

4-4 Study Guide and Intervention (continued)

Parallel and Perpendicular Lines

Perpendicular Lines Two nonvertical lines are **perpendicular** if their slopes are negative reciprocals of each other. Vertical and horizontal lines are perpendicular.

Example Write an equation in slope-intercept form for the line that passes through $(-4, 2)$ and is perpendicular to the graph of $2x - 3y = 9$.

Find the slope of $2x - 3y = 9$.

$2x - 3y = 9$ Original equation

$-3y = -2x + 9$ Subtract 2x from each side.

$y = \dfrac{2}{3}x - 3$ Divide each side by −3.

The slope of $y = \dfrac{2}{3}x - 3$ is $\dfrac{2}{3}$. So, the slope of the line passing through $(-4, 2)$ that is perpendicular to this line is the negative reciprocal of $\dfrac{2}{3}$, or $-\dfrac{3}{2}$.

Use the point-slope form to find the equation.

$y - y_1 = m(x - x_1)$ Point-slope form

$y - 2 = -\dfrac{3}{2}(x - (-4))$ $m = -\dfrac{3}{2}; (x_1, y_1) = (-4, 2)$

$y - 2 = -\dfrac{3}{2}(x + 4)$ Simplify.

$y - 2 = -\dfrac{3}{2}x - 6$ Distributive Property

$y = -\dfrac{3}{2}x - 4$ Slope-intercept form

Exercises

1. **ARCHITECTURE** On the architect's plans for a new high school, a wall represented by $\overline{MN}$ has endpoints $M(-3, -1)$ and $N(2, 1)$. A wall represented by $\overline{PQ}$ has endpoints $P(4, -4)$ and $Q(-2, 11)$. Are the walls perpendicular? Explain.

Determine whether the graphs of the following equations are *parallel* or *perpendicular*.

2. $2x + y = -7, x - 2y = -4, 4x - y = 5$

3. $y = 3x, 6x - 2y = 7, 3y = 9x - 1$

Write an equation in slope-intercept form for the line that passes through the given point and is perpendicular to the graph of each equation.

4. $(4, 2), y = \dfrac{1}{2}x + 1$ 5. $(2, -3), y = -\dfrac{2}{3}x + 4$ 6. $(6, 4), y = 7x + 1$

7. $(-8, -7), y = -x - 8$ 8. $(6, -2), y = -3x - 6$ 9. $(-5, -1), y = \dfrac{5}{2}x - 3$

4-5 Study Guide and Intervention

Scatter Plots and Lines of Fit

Investigate Relationships Using Scatter Plots A **scatter plot** is a graph in which two sets of data are plotted as ordered pairs in a coordinate plane. If y increases as x increases, there is a **positive correlation** between x and y. If y decreases as x increases, there is a **negative correlation** between x and y. If x and y are not related, there is **no correlation**.

Example **EARNINGS** The graph at the right shows the amount of money Carmen earned each week and the amount she deposited in her savings account that same week. Determine whether the graph shows a *positive correlation*, a *negative correlation*, or *no correlation*. If there is a positive or negative correlation, describe its meaning in the situation.

The graph shows a positive correlation. The more Carmen earns, the more she saves.

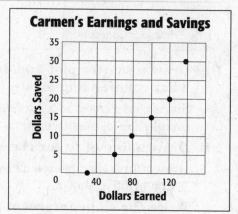

Carmen's Earnings and Savings

Exercises

Determine whether each graph shows a *positive correlation*, a *negative correlation*, or *no correlation*. If there is a positive or negative correlation, describe its meaning in the situation.

1. **Average Weekly Work Hours in U.S.**

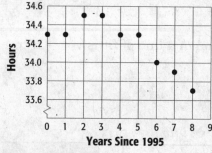

Source: *The World Almanac*

2. **Average Jogging Speed**

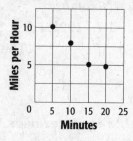

3. **Average U.S. Hourly Earnings**

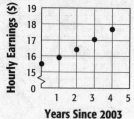

Source: U.S. Dept. of Labor

4. **U.S. Imports from Mexico**

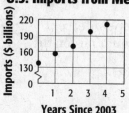

Source: U.S. Census Bureau

4-5 Study Guide and Intervention *(continued)*

Scatter Plots and Lines of Fit

Use Lines of Fit

Example The table shows the number of students per computer in Easton High School for certain school years from 1996 to 2008.

Year	1996	1998	2000	2002	2004	2006	2008
Students per Computer	22	18	14	10	6.1	5.4	4.9

a. Draw a scatter plot and determine what relationship exists, if any.

Since y decreases as x increases, the correlation is negative.

b. Draw a line of fit for the scatter plot.

Draw a line that passes close to most of the points. A line of fit is shown.

c. Write the slope-intercept form of an equation for the line of fit.

The line of fit shown passes through (1999, 16) and (2005, 5.7). Find the slope.

$$m = \frac{5.7 - 16}{2005 - 1999}$$

$$m = -1.7$$

Find b in $y = -1.7x + b$.

$$16 = -1.7 \cdot 1993 + b$$

$$3404 = b$$ Therefore, an equation of a line of fit is $y = -1.7x + 3404$.

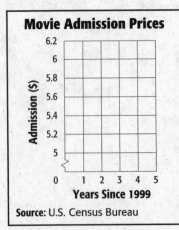

Students per Computer in Easton High School

Exercises

Refer to the table for Exercises 1–3.

1. Draw a scatter plot.

2. Draw a line of fit for the data.

3. Write the slope-intercept form of an equation for the line of fit.

Movie Admission Prices

Source: U.S. Census Bureau

Years Since 1999	Admission (dollars)
0	$5.08
1	$5.39
2	$5.66
3	$5.81
4	$6.03

4-6 Study Guide and Intervention
Regression and Median-Fit Lines

Equations of Best-Fit Lines Many graphing calculators utilize an algorithm called **linear regression** to find a precise line of fit called the **best-fit line**. The calculator computes the data, writes an equation, and gives you the **correlation coefficent**, a measure of how closely the equation models the data.

Example **GAS PRICES** The table shows the price of a gallon of regular gasoline at a station in Los Angeles, California on January 1 of various years.

Year	2005	2006	2007	2008	2009	2010
Average Price	$1.47	$1.82	$2.15	$2.49	$2.83	$3.04

Source: U.S. Department of Energy

a. **Use a graphing calculator to write an equation for the best-fit line for that data.** Enter the data by pressing [STAT] and selecting the Edit option. Let the year 2005 be represented by 0. Enter the years since 2005 into List 1 (L1). Enter the average price into List 2 (L2).

Then, perform the linear regression by pressing [STAT] and selecting the CALC option. Scroll down to LinReg (ax+b) and press [ENTER]. The best-fit equation for the regression is shown to be $y = 0.321x + 1.499$.

b. **Name the correlation coefficient.** The correlation coefficient is the value shown for r on the calculator screen. The correlation coefficient is about 0.998.

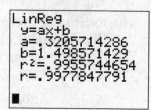

Exercises

Write an equation of the regression line for the data in each table below. Then find the correlation coefficient.

1. **OLYMPICS** Below is a table showing the number of gold medals won by the United States at the Winter Olympics during various years.

Year	1992	1994	1998	2002	2006	2010
Gold Medals	5	6	6	10	9	9

Source: International Olympic Committee

2. **INTEREST RATES** Below is a table showing the U.S. Federal Reserve's prime interest rate on January 1 of various years.

Year	2006	2007	2008	2009	2010
Prime Rate (percent)	7.25	8.25	7.25	3.25	3.25

Source: Federal Reserve Board

4-6 Study Guide and Intervention (continued)

Regression and Median-Fit Lines

Equations of Median-Fit Lines A graphing calculator can also find another type of best-fit line called the **median-fit line**, which is found using the medians of the coordinates of the data points.

Example **ELECTIONS** The table shows the total number of people in millions who voted in the U.S. Presidential election in the given years.

Year	1980	1984	1988	1992	1996	2004	2008
Voters	86.5	92.7	91.6	104.4	96.3	122.3	131.3

Source: George Mason University

a. Find an equation for the median-fit line. Enter the data by pressing ⃞STAT and selecting the Edit option. Let the year 1980 be represented by 0. Enter the years since 1980 into List 1 (L1). Enter the number of voters into List 2 (L2).

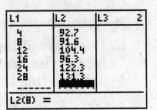

Then, press ⃞STAT and select the CALC option. Scroll down to Med-Med option and press ⃞ENTER. The value of a is the slope, and the value of b is the y-intercept.

The equation for the median-fit line is $y = 1.55x + 83.57$.

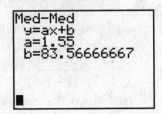

b. Estimate the number of people who voted in the 2000 U.S. Presidential election. Graph the best-fit line. Then use the ⃞TRACE feature and the arrow keys until you find a point where $x = 20$.

When $x = 20$, $y \approx 115$. Therefore, about 115 million people voted in the 2000 U.S. Presidential election.

Exercises

Write an equation of the regression line for the data in each table below. Then find the correlation coefficient.

1. POPULATION GROWTH Below is a table showing the estimated population of Arizona in millions on July 1st of various years.

Year	2001	2002	2003	2004	2005	2006
Population	5.30	5.44	5.58	5.74	5.94	6.17

Source: U.S. Census Bureau

a. Find an equation for the median-fit line.

b. Predict the population of Arizona in 2009.

2. ENROLLMENT Below is a table showing the number of students enrolled at Happy Days Preschool in the given years.

Year	2002	2004	2006	2008	2010
Students	130	168	184	201	234

a. Find an equation for the median-fit line.

b. Estimate how many students were enrolled in 2007.

4-7 Study Guide and Intervention

Inverse Linear Functions

Inverse Relations An **inverse relation** is the set of ordered pairs obtained by exchanging the x-coordinates with the y-coordinates of each ordered pair. The domain of a relation becomes the range of its inverse, and the range of the relation becomes the domain of its inverse.

Example **Find and graph the inverse of the relation represented by line a.**

The graph of the relation passes through $(-2, -10)$, $(-1, -7)$, $(0, -4)$, $(1, -1)$, $(2, 2)$, $(3, 5)$, and $(4, 8)$.

To find the inverse, exchange the coordinates of the ordered pairs.

The graph of the inverse passes through the points $(-10, -2)$, $(-7, -1)$, $(-4, 0)$, $(-1, 1)$, $(2, 2)$, $(5, 3)$, and $(8, 4)$. Graph these points and then draw the line that passes through them.

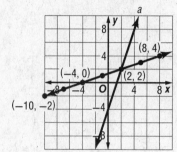

Exercises

Find the inverse of each relation.

1. $\{(4, 7), (6, 2), (9, -1), (11, 3)\}$

2. $\{(-5, -9), (-4, -6), (-2, -4), (0, -3)\}$

3.

x	y
−8	−15
−2	−11
1	−8
5	1
11	8

4.

x	y
−8	3
−2	9
2	13
6	18
8	19

5.

x	y
−6	14
−5	11
−4	8
−3	5
−2	2

Graph the inverse of each relation.

6.

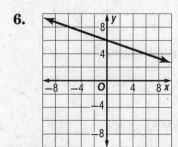

7.

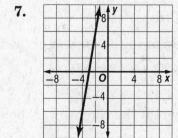

8.

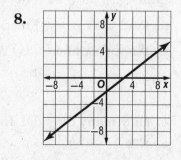

4-7   Study Guide and Intervention (continued)
Inverse Linear Functions

Inverse Functions A linear relation that is described by a function has an **inverse function** that can generate ordered pairs of the inverse relation. The inverse of the linear function $f(x)$ can be written as $f^{-1}(x)$ and is read *f of x inverse* or *the inverse of f of x*.

> **Example**   Find the inverse of $f(x) = \frac{3}{4}x + 6$.

Step 1 $f(x) = \frac{3}{4}x + 6$ Original equation

   $y = \frac{3}{4}x + 6$ Replace $f(x)$ with y.

Step 2 $x = \frac{3}{4}y + 6$ Interchange y and x.

Step 3 $x - 6 = \frac{3}{4}y$ Subtract 6 from each side.

   $\frac{4}{3}(x - 6) = y$ Multiply each side by $\frac{4}{3}$.

Step 4 $\frac{4}{3}(x - 6) = f^{-1}(x)$ Replace y with $f^{-1}(x)$.

The inverse of $f(x) = \frac{3}{4}x + 6$ is $f^{-1}(x) = \frac{4}{3}(x - 6)$ or $f^{-1}(x) = \frac{4}{3}x - 8$.

Exercises

Find the inverse of each function.

1. $f(x) = 4x - 3$ **2.** $f(x) = -3x + 7$ **3.** $f(x) = \frac{3}{2}x - 8$

4. $f(x) = 16 - \frac{1}{3}x$ **5.** $f(x) = 3(x - 5)$ **6.** $f(x) = -15 - \frac{2}{5}x$

7. TOOLS Jimmy rents a chainsaw from the department store to work on his yard. The total cost $C(x)$ in dollars is given by $C(x) = 9.99 + 3.00x$, where x is the number of days he rents the chainsaw.

   a. Find the inverse function $C^{-1}(x)$.

   b. What do x and $C^{-1}(x)$ represent in the context of the inverse function?

   c. How many days did Jimmy rent the chainsaw if the total cost was $27.99?

5-1   Study Guide and Intervention

Solving Inequalities by Addition and Subtraction

Solve Inequalities by Addition Addition can be used to solve inequalities. If any number is added to each side of a true inequality, the resulting inequality is also true.

Addition Property of Inequalities	For all numbers a, b, and c, if $a > b$, then $a + c > b + c$, and if $a < b$, then $a + c < b + c$.

The property is also true when $>$ and $<$ are replaced with $\geq$ and $\leq$.

Example 1 Solve $x - 8 \leq -6$. Then graph the solution.

$$x - 8 \leq -6 \qquad \text{Original inequality}$$
$$x - 8 + 8 \leq -6 + 8 \qquad \text{Add 8 to each side.}$$
$$x \leq 2 \qquad \text{Simplify.}$$

The solution in set-builder notation is $\{x \mid x \leq 2\}$.

Number line graph:

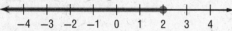

Example 2 Solve $4 - 2a > -a$. Then graph the solution.

$$4 - 2a > -a \qquad \text{Original inequality}$$
$$4 - 2a + 2a > -a + 2a \qquad \text{Add 2a to each side.}$$
$$4 > a \qquad \text{Simplify.}$$
$$a < 4 \qquad 4 > a \text{ is the same as } a < 4.$$

The solution in set-builder notation is $\{a \mid a < 4\}$.

Number line graph:

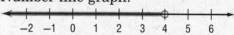

Exercises

Solve each inequality. Check your solution, and then graph it on a number line.

1. $t - 12 \geq 16$

26 27 28 29 30 31 32 33 34

2. $n - 12 < 6$

12 13 14 15 16 17 18 19 20

3. $6 \leq g - 3$

7 8 9 10 11 12 13 14 15

4. $n - 8 < -13$

-10 -9 -8 -7 -6 -5 -4 -3 -2

5. $-12 > -12 + y$

-4 -3 -2 -1 0 1 2 3 4

6. $-6 > m - 8$

-4 -3 -2 -1 0 1 2 3 4

Solve each inequality. Check your solution.

7. $-3x \leq 8 - 4x$

8. $0.6n \geq 12 - 0.4n$

9. $-8k - 12 < -9k$

10. $-y - 10 > 15 - 2y$

11. $z - \frac{1}{3} \leq \frac{4}{3}$

12. $-2b > -4 - 3b$

Define a variable, write an inequality, and solve each problem. Check your solution.

13. A number decreased by 4 is less than 14.

14. The difference of two numbers is more than 12, and one of the numbers is 3.

15. Forty is no greater than the difference of a number and 2.

5-1 Study Guide and Intervention *(continued)*

Solving Inequalities by Addition and Subtraction

Solve Inequalities by Subtraction Subtraction can be used to solve inequalities. If any number is subtracted from each side of a true inequality, the resulting inequality is also true.

Subtraction Property of Inequalities	For all numbers a, b, and c, if $a > b$, then $a - c > b - c$, and if $a < b$, then $a - c < b - c$.

The property is also true when $>$ and $<$ are replaced with $\geq$ and $\leq$.

Example **Solve $3a + 5 > 4 + 2a$. Then graph it on a number line.**

$$3a + 5 > 4 + 2a \qquad \text{Original inequality}$$
$$3a + 5 - 2a > 4 + 2a - 2a \qquad \text{Subtract } 2a \text{ from each side.}$$
$$a + 5 > 4 \qquad \text{Simplify.}$$
$$a + 5 - 5 > 4 - 5 \qquad \text{Subtract 5 from each side.}$$
$$a > -1 \qquad \text{Simplify.}$$

The solution is $\{a | a > -1\}$.

Number line graph:

```
←――+――+――+――○――+――+――+――+――→
   -4  -3  -2  -1   0   1   2   3   4
```

Exercises

Solve each inequality. Check your solution, and then graph it on a number line.

1. $t + 12 \geq 8$

2. $n + 12 > -12$

3. $16 \leq h + 9$

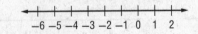

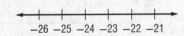

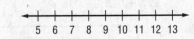

4. $y + 4 > -2$

5. $3r + 6 > 4r$

6. $\frac{3}{2}q - 5 \geq \frac{1}{2}q$

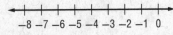

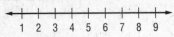

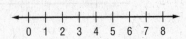

Solve each inequality. Check your solution.

7. $4p \geq 3p + 0.7$

8. $r + \frac{1}{4} > \frac{3}{8}$

9. $9k + 12 > 8k$

10. $-1.2 > 2.4 + y$

11. $4y < 5y + 14$

12. $3n + 17 < 4n$

Define a variable, write an inequality, and solve each problem. Check your solution.

13. The sum of a number and 8 is less than 12.

14. The sum of two numbers is at most 6, and one of the numbers is -2.

15. The sum of a number and 6 is greater than or equal to -4.

5-2 Study Guide and Intervention

Solving Inequalities by Multiplication and Division

Solve Inequalities by Multiplication If each side of an inequality is multiplied by the same positive number, the resulting inequality is also true. However, if each side of an inequality is multiplied by the same negative number, the direction of the inequality must be reversed for the resulting inequality to be true.

Multiplication Property of Inequalities	For all numbers a, b, and c, with $c \neq 0$,
	1. if c is positive and $a > b$, then $ac > bc$;
	   if c is positive and $a < b$, then $ac < bc$;
	2. if c is negative and $a > b$, then $ac < bc$;
	   if c is negative and $a < b$, then $ac > bc$.

The property is also true when $>$ and $<$ are replaced with $\geq$ and $\leq$.

Example 1 Solve $-\dfrac{y}{8} \leq 12$.

$-\dfrac{y}{8} \geq 12$ Original inequality

$(-8)\left(-\dfrac{y}{8}\right) \leq (-8)12$ Multiply each side by −8; change $\geq$ to $\leq$.

$y \leq -96$ Simplify.

The solution is $\{y \mid y \leq -96\}$.

Example 2 Solve $\dfrac{3}{4}k < 15$.

$\dfrac{3}{4}k < 15$ Original inequality

$\left(\dfrac{4}{3}\right)\dfrac{3}{4}k < \left(\dfrac{4}{3}\right)15$ Multiply each side by $\dfrac{4}{3}$.

$k < 20$ Simplify.

The solution is $\{k \mid k < 20\}$.

Exercises

Solve each inequality. Check your solution.

1. $\dfrac{y}{6} \leq 2$ **2.** $-\dfrac{n}{50} > 22$ **3.** $\dfrac{3}{5}h \geq -3$ **4.** $-\dfrac{p}{6} < -6$

5. $\dfrac{1}{4}n \geq 10$ **6.** $-\dfrac{2}{3}b < \dfrac{1}{3}$ **7.** $\dfrac{3m}{5} < -\dfrac{3}{20}$ **8.** $-2.51 \leq -\dfrac{2h}{4}$

9. $\dfrac{g}{5} \geq -2$ **10.** $-\dfrac{3}{4} > -\dfrac{9p}{5}$ **11.** $\dfrac{n}{10} \geq 5.4$ **12.** $\dfrac{2a}{7} \geq -6$

Define a variable, write an inequality, and solve each problem. Check your solution.

13. Half of a number is at least 14.

14. The opposite of one-third a number is greater than 9.

15. One fifth of a number is at most 30.

5-2 Study Guide and Intervention (continued)

Solving Inequalities by Multiplication and Division

Solve Inequalities by Division If each side of a true inequality is divided by the same positive number, the resulting inequality is also true. However, if each side of an inequality is divided by the same negative number, the direction of the inequality symbol must be reversed for the resulting inequality to be true.

Division Property of Inequalities	For all numbers a, b, and c with $c \neq 0$, **1.** if c is positive and $a > b$, then $\frac{a}{c} > \frac{b}{c}$; if c is positive and $a < b$, then $\frac{a}{c} < \frac{b}{c}$; **2.** if c is negative and $a > b$, then $\frac{a}{c} < \frac{b}{c}$; if c is negative and $a < b$, then $\frac{a}{c} > \frac{b}{c}$.

The property is also true when $>$ and $<$ are replaced with $\geq$ and $\leq$.

Example **Solve $-12y \geq 48$.**

$-12y \geq 48$ Original inequality

$\dfrac{-12y}{-12} \leq \dfrac{48}{-12}$ Divide each side by -12 and change $\geq$ to $\leq$.

$y \leq -4$ Simplify.

The solution is $\{y \mid y \leq -4\}$.

Exercises

Solve each inequality. Check your solution.

1. $25g \geq -100$ **2.** $-2x \geq 9$ **3.** $-5c > 2$ **4.** $-8m < -64$

5. $-6k < \dfrac{1}{5}$ **6.** $18 < -3b$ **7.** $30 < -3n$ **8.** $-0.24 < 0.6w$

9. $25 \geq -2m$ **10.** $-30 > -5p$ **11.** $-2n \geq 6.2$ **12.** $35 < 0.05h$

13. $-40 > 10h$ **14.** $-\dfrac{2}{3n} \geq 6$ **15.** $-3 < \dfrac{p}{4}$ **16.** $4 > \dfrac{-x}{2}$

Define a variable, write an inequality, and solve each problem. Then check your solution.

17. Four times a number is no more than 108.

18. The opposite of three times a number is greater than 12.

19. Negative five times a number is at most 100.

5-3   Study Guide and Intervention

Solving Multi-Step Inequalities

Solve Multi-Step Inequalities To solve linear inequalities involving more than one operation, undo the operations in reverse of the order of operations, just as you would solve an equation with more than one operation.

Example 1	Solve $6x - 4 \leq 2x + 12$.

$6x - 4 \leq 2x + 12$	Original inequality
$6x - 4 - 2x \leq 2x + 12 - 2x$	Subtract 2x from each side.
$4x - 4 \leq 12$	Simplify.
$4x - 4 + 4 \leq 12 + 4$	Add 4 to each side.
$4x \leq 16$	Simplify.
$\dfrac{4x}{4} \leq \dfrac{16}{4}$	Divide each side by 4.
$x \leq 4$	Simplify.

The solution is $\{x \mid x \leq 4\}$.

Example 2	Solve $3a - 15 > 4 + 5a$.

$3a - 15 > 4 + 5a$	Original inequality
$3a - 15 - 5a > 4 + 5a - 5a$	Subtract 5a from each side.
$-2a - 15 > 4$	Simplify.
$-2a - 15 + 15 > 4 + 15$	Add 15 to each side.
$-2a > 19$	Simplify.
$\dfrac{-2a}{-2} < \dfrac{19}{-2}$	Divide each side by −2 and change > to <.
$a < -9\dfrac{1}{2}$	Simplify.

The solution is $\left\{a \mid a < -9\dfrac{1}{2}\right\}$.

Exercises

Solve each inequality. Check your solution.

1. $11y + 13 \geq -1$

2. $8n - 10 < 6 - 2n$

3. $\dfrac{q}{7} + 1 > -5$

4. $6n + 12 < 8 + 8n$

5. $-12 - d > -12 + 4d$

6. $5r - 6 > 8r - 18$

7. $\dfrac{-3x + 6}{2} \leq 12$

8. $7.3y - 14.4 > 4.9y$

9. $-8m - 3 < 18 - m$

10. $-4y - 10 > 19 - 2y$

11. $9n - 24n + 45 > 0$

12. $\dfrac{4x - 2}{5} \geq -4$

Define a variable, write an inequality, and solve each problem. Check your solution.

13. Negative three times a number plus four is no more than the number minus eight.

14. One fourth of a number decreased by three is at least two.

15. The sum of twelve and a number is no greater than the sum of twice the number and −8.

5-3 Study Guide and Intervention *(continued)*

Solving Multi-Step Inequalities

Solve Inequalities Involving the Distributive Property When solving inequalities that contain grouping symbols, first use the Distributive Property to remove the grouping symbols. Then undo the operations in reverse of the order of operations, just as you would solve an equation with more than one operation.

Example Solve $3a - 2(6a - 4) > 4 - (4a + 6)$.

$3a - 2(6a - 4) > 4 - (4a + 6)$	Original inequality
$3a - 12a + 8 > 4 - 4a - 6$	Distributive Property
$-9a + 8 > -2 - 4a$	Combine like terms.
$-9a + 8 + 4a > -2 - 4a + 4a$	Add 4a to each side.
$-5a + 8 > -2$	Combine like terms.
$-5a + 8 - 8 > -2 - 8$	Subtract 8 from each side.
$-5a > -10$	Simplify.
$a < 2$	Divide each side by −5 and change > to <.

The solution in set-builder notation is $\{a \mid a < 2\}$.

Exercises

Solve each inequality. Check your solution.

1. $2(t + 3) \geq 16$

2. $3(d - 2) - 2d > 16$

3. $4h - 8 < 2(h - 1)$

4. $6y + 10 > 8 - (y + 14)$

5. $4.6(x - 3.4) > 5.1x$

6. $-5x - (2x + 3) \geq 1$

7. $3(2y - 4) - 2(y + 1) > 10$

8. $8 - 2(b + 1) < 12 - 3b$

9. $-2(k - 1) > 8(1 + k)$

10. $0.3(y - 2) > 0.4(1 + y)$

11. $m + 17 \leq -(4m - 13)$

12. $3n + 8 \leq 2(n - 4) - 2(1 - n)$

13. $2(y - 2) > -4 + 2y$

14. $k - 17 \leq -(17 - k)$

15. $n - 4 \leq -3(2 + n)$

Define a variable, write an inequality, and solve each problem. Check your solution.

16. Twice the sum of a number and 4 is less than 12.

17. Three times the sum of a number and six is greater than four times the number decreased by two.

18. Twice the difference of a number and four is less than the sum of the number and five.

5-4 Study Guide and Intervention

Solving Compound Inequalities

Inequalities Containing *and* A compound inequality containing *and* is true only if both inequalities are true. The graph of a compound inequality containing *and* is the **intersection** of the graphs of the two inequalities. Every solution of the compound inequality must be a solution of both inequalities.

Example 1 Graph the solution set of $x < 2$ and $x \geq -1$.

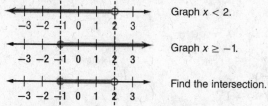

Graph $x < 2$.

Graph $x \geq -1$.

Find the intersection.

The solution set is $\{x \mid -1 \leq x < 2\}$.

Example 2 Solve $-1 < x + 2 < 3$. Then graph the solution set.

$$-1 < x + 2 \qquad \text{and} \qquad x + 2 < 3$$
$$-1 - 2 < x + 2 - 2 \qquad\qquad x + 2 - 2 < 3 - 2$$
$$-3 < x \qquad\qquad\qquad x < 1$$

Graph $x > -3$.

Graph $x < 1$.

Find the intersection.

The solution set is $\{x \mid -3 < x < 1\}$.

Exercises

Graph the solution set of each compound inequality.

1. $b > -1$ and $b \leq 3$

2. $2 \geq q \geq -5$

3. $x > -3$ and $x \leq 4$

4. $-2 \leq p < 4$

5. $-3 < d$ and $d < 2$

6. $-1 \leq p \leq 3$

Solve each compound inequality. Then graph the solution set.

7. $4 < w + 3 \leq 5$

8. $-3 \leq p - 5 < 2$

9. $-4 < x + 2 \leq -2$

10. $y - 1 < 2$ and $y + 2 \geq 1$

11. $n - 2 > -3$ and $n + 4 < 6$

12. $d - 3 < 6d + 12 < 2d + 32$

5-4 Study Guide and Intervention (continued)

Solving Compound Inequalities

Inequalities Containing *or* A compound inequality containing *or* is true if one or both of the inequalities are true. The graph of a compound inequality containing *or* is the **union** of the graphs of the two inequalities. The union can be found by graphing both inequalities on the same number line. A solution of the compound inequality is a solution of either inequality, not necessarily both.

Example **Solve $2a + 1 < 11$ or $a > 3a + 2$. Then graph the solution set.**

$$2a + 1 < 11$$
$$2a + 1 - 1 < 11 - 1$$
$$2a < 10$$
$$\frac{2a}{2} < \frac{10}{2}$$
$$a < 5$$

or

$$a > 3a + 2$$
$$a - 3a > 3a - 3a + 2$$
$$-2a > 2$$
$$\frac{-2a}{-2} < \frac{2}{-2}$$
$$a < -1$$

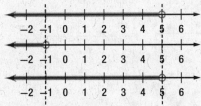

Graph $a < 5$.

Graph $a < -1$.

Find the union.

The solution set is $\{a \mid a < 5\}$.

Exercises

Graph the solution set of each compound inequality.

1. $b > 2$ or $b \le -3$

-4 -3 -2 -1 0 1 2 3 4

2. $3 \ge q$ or $q \le 1$

-4 -3 -2 -1 0 1 2 3 4

3. $y \le -4$ or $y > 0$

-5 -4 -3 -2 -1 0 1 2 3

4. $4 \le p$ or $p < 8$

-2 -1 0 1 2 3 4 5 6

5. $-3 < d$ or $d < 2$

-4 -3 -2 -1 0 1 2 3 4

6. $-2 \le x$ or $3 \le x$

-4 -3 -2 -1 0 1 2 3 4

Solve each compound inequality. Then graph the solution set.

7. $3 < 3w$ or $3w \ge 9$

-4 -3 -2 -1 0 1 2 3 4

8. $-3p + 1 \le -11$ or $p < 2$

0 1 2 3 4 5 6 7 8

9. $2x + 4 \le 6$ or $x \ge 2x - 4$

-2 -1 0 1 2 3 4 5 6

10. $2y + 2 < 12$ or $y - 3 \ge 2y$

0 1 2 3 4 5 6 7 8

11. $\frac{1}{2}n > -2$ or $2n - 2 < 6 + n$

-4 -3 -2 -1 0 1 2 3 4

12. $3a + 2 \ge 5$ or $7 + 3a < 2a + 6$

-4 -3 -2 -1 0 1 2 3 4

5-5 Study Guide and Intervention

Inequalities Involving Absolute Value

Inequalities Involving Absolute Value (<) When solving inequalities that involve absolute value, there are two cases to consider for inequalities involving < (or ≤).

If $|x| < n$, then $x > -n$ and $x < n$.

Remember that inequalities with *and* are related to intersections.

Example Solve $|3a + 4| < 10$. Then graph the solution set.

Write $|3a + 4| < 10$ as $3a + 4 < 10$ and $3a + 4 > -10$.

$3a + 4 < 10$	and	$3a + 4 > -10$
$3a + 4 - 4 < 10 - 4$		$3a + 4 - 4 > -10 - 4$
$3a < 6$		$3a > -14$
$\frac{3a}{3} < \frac{6}{3}$		$\frac{3a}{3} > \frac{-14}{3}$
$a < 2$		$a > -4\frac{2}{3}$

Now graph the solution set.

The solution set is $\left\{ a \mid -4\frac{2}{3} < a < 2 \right\}$.

Exercises

Solve each inequality. Then graph the solution set.

1. $|y| < 3$

2. $|x - 4| < 4$

3. $|y + 3| \le 2$

4. $|b + 2| \le 3$

5. $|w - 2| \le 5$

6. $|t + 2| \le 4$

7. $|2x| \le 8$

8. $|5y - 2| \le 7$

9. $|p - 0.2| < 0.5$

5-5 Study Guide and Intervention (continued)

Inequalities Involving Absolute Value

Solve Absolute Value Inequalities (>) When solving inequalities that involve absolute value, there are two cases to consider for inequalities involving > (or ≥). Remember that inequalities with *or* are related to unions.

Example Solve $|2b + 9| > 5$. Then graph the solution set.

Write $|2b + 9| > 5$ as $|2b + 9| > 5$ or $|2b + 9| < -5$.

$$2b + 9 > 5 \qquad \text{or} \qquad 2b + 9 < -5$$
$$2b + 9 - 9 > 5 - 9 \qquad\qquad 2b + 9 - 9 < -5 - 9$$
$$2b > -4 \qquad\qquad\qquad 2b < -14$$
$$\frac{2b}{2} > \frac{-4}{2} \qquad\qquad\qquad \frac{2b}{2} < \frac{-14}{2}$$
$$b > -2 \qquad\qquad\qquad b < -7$$

Now graph the solution set.

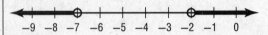

The solution set is $\{b \mid b > -2 \text{ or } b < -7\}$.

Exercises

Solve each inequality. Then graph the solution set.

1. $|c - 2| > 6$

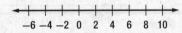

2. $|x - 3| > 0$

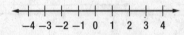

3. $|3f + 10| \geq 4$

4. $|x| \geq 2$

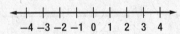

5. $|x| \geq 3$

6. $|2x + 1| \geq -2$

7. $|2d - 1| \geq 4$

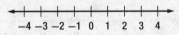

8. $|3 - (x - 1)| \geq 8$

9. $|3r + 2| > -5$

5-6 Study Guide and Intervention

Graphing Inequalities in Two Variables

Graph Linear Inequalities The solution set of an inequality that involves two variables is graphed by graphing a related linear equation that forms a boundary of a **half-plane**. The graph of the ordered pairs that make up the solution set of the inequality fill a region of the coordinate plane on one side of the half-plane.

> **Example** **Graph $y \leq -3x - 2$.**
>
> Graph $y = -3x - 2$.
>
> Since $y \leq -3x - 2$ is the same as $y < -3x - 2$ and $y = -3x - 2$, the boundary is included in the solution set and the graph should be drawn as a solid line.
>
> Select a point in each half plane and test it. Choose $(0, 0)$ and $(-2, -2)$.
>
> $y \leq -3x - 2$ $y \leq -3x - 2$
>
> $0 \leq -3(0) - 2$ $-2 \leq -3(-2) - 2$
>
> $0 \leq -2$ is false. $-2 \leq 6 - 2$
>
> $-2 \leq 4$ is true.
>
> The half-plane that contains $(-2, -2)$ contains the solution. Shade that half-plane.

Exercises

Graph each inequality.

1. $y < 4$

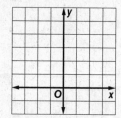

2. $x \geq 1$

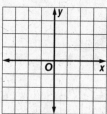

3. $3x \leq y$

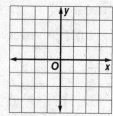

4. $-x > y$

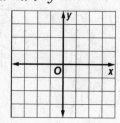

5. $x - y \geq 1$

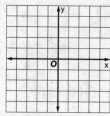

6. $2x - 3y \leq 6$

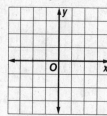

7. $y < -\frac{1}{2}x - 3$

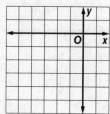

8. $4x - 3y < 6$

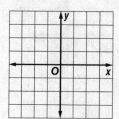

9. $3x + 6y \geq 12$

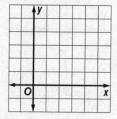

5-6 Study Guide and Intervention (continued)

Graphing Inequalities in Two Variables

Solve Linear Inequalities We can use a coordinate plane to solve inequalities with one variable.

Example Use a graph to solve $2x + 2 > -1$.

Step 1 First graph the boundary, which is the related function. Replace the inequality sign with an equals sign, and get 0 on a side by itself.

$$2x + 2 > -1 \qquad \text{Original inequality}$$
$$2x + 2 = -1 \qquad \text{Change} < \text{to} = .$$
$$2x + 2 + 1 = -1 + 1 \qquad \text{Add 1 to each side.}$$
$$2x + 3 = 0 \qquad \text{Simplify.}$$

Graph $2x + 3 = y$ as a dashed line.

Step 2 Choose $(0, 0)$ as a test point, substituting these values into the original inequality give us $3 > -5$.

Step 3 Because this statement is true, shade the half plane containing the point $(0, 0)$.

Notice that the x-intercept of the graph is at $-1\frac{1}{2}$. Because the half-plane to the right of the x-intercept is shaded, the solution is $x > -1\frac{1}{2}$.

Exercises

Use a graph to solve each inequality.

1. $x + 7 \le 5$

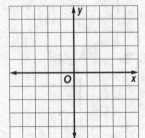

2. $x - 2 > 2$

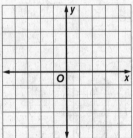

3. $-x + 1 < -3$

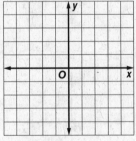

4. $-x - 7 \ge -6$

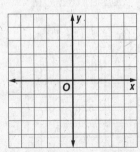

5. $3x - 20 < -17$

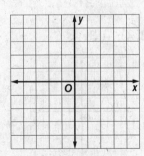

6. $-2x + 11 \ge 15$

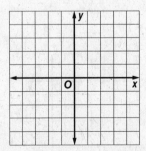

6-1　Study Guide and Intervention

Graphing Systems of Equations

Possible Number of Solutions Two or more linear equations involving the same variables form a **system of equations**. A solution of the system of equations is an ordered pair of numbers that satisfies both equations. The table below summarizes information about systems of linear equations.

Graph of a System	intersecting lines	same line	parallel lines
	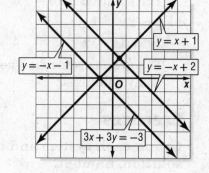		
Number of Solutions	exactly one solution	infinitely many solutions	no solution
Terminology	consistent and independent	consistent and dependent	inconsistent

Example　Use the graph at the right to determine whether each system is *consistent* or *inconsistent* and if it is *independent* or *dependent*.

a. $y = -x + 2$
　　$y = x + 1$

Since the graphs of $y = -x + 2$ and $y = x + 1$ intersect, there is one solution. Therefore, the system is consistent and independent.

b. $y = -x + 2$
　　$3x + 3y = -3$

Since the graphs of $y = -x + 2$ and $3x + 3y = -3$ are parallel, there are no solutions. Therefore, the system is inconsistent.

c. $3x + 3y = -3$
　　$y = -x - 1$

Since the graphs of $3x + 3y = -3$ and $y = -x - 1$ coincide, there are infinitely many solutions. Therefore, the system is consistent and dependent.

Exercises

Use the graph at the right to determine whether each system is *consistent* or *inconsistent* and if it is *independent* or *dependent*.

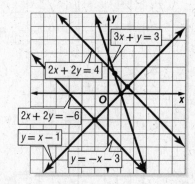

1. $y = -x - 3$
　　$y = x - 1$

2. $2x + 2y = -6$
　　$y = -x - 3$

3. $y = -x - 3$
　　$2x + 2y = 4$

4. $2x + 2y = -6$
　　$3x + y = 3$

6-1 Study Guide and Intervention (continued)

Graphing Systems of Equations

Solve by Graphing One method of solving a system of equations is to graph the equations on the same coordinate plane.

Example Graph each system and determine the number of solutions that it has. If it has one solution, name it.

a. $x + y = 2$
$x - y = 4$

The graphs intersect. Therefore, there is one solution. The point $(3, -1)$ seems to lie on both lines. Check this estimate by replacing x with 3 and y with -1 in each equation.

$x + y = 2$
$3 + (-1) = 2$ ✔
$x - y = 4$
$3 - (-1) = 3 + 1$ or 4 ✔
The solution is $(3, -1)$.

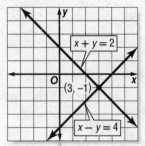

b. $y = 2x + 1$
$2y = 4x + 2$

The graphs coincide. Therefore there are infinitely many solutions.

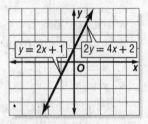

Exercises

Graph each system and determine the number of solutions it has. If it has one solution, name it.

1. $y = -2$
$3x - y = -1$

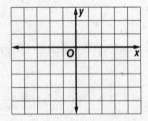

2. $x = 2$
$2x + y = 1$

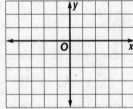

3. $y = \frac{1}{2}x$
$x + y = 3$

4. $2x + y = 6$
$2x - y = -2$

5. $3x + 2y = 6$
$3x + 2y = -4$

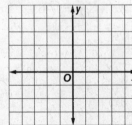

6. $2y = -4x + 4$
$y = -2x + 2$

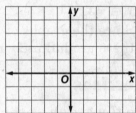

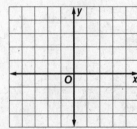

6-2 Study Guide and Intervention

Substitution

Solve by Substitution One method of solving systems of equations is **substitution**.

Example 1 Use substitution to solve the system of equations.

$y = 2x$
$4x - y = -4$

Substitute $2x$ for y in the second equation.

$4x - y = -4$	Second equation
$4x - 2x = -4$	$y = 2x$
$2x = -4$	Combine like terms.
$x = -2$	Divide each side by 2 and simplify.

Use $y = 2x$ to find the value of y.

$y = 2x$	First equation
$y = 2(-2)$	$x = -2$
$y = -4$	Simplify.

The solution is $(-2, -4)$.

Example 2 Solve for one variable, then substitute.

$x + 3y = 7$
$2x - 4y = -6$

Solve the first equation for x since the coefficient of x is 1.

$x + 3y = 7$	First equation
$x + 3y - 3y = 7 - 3y$	Subtract 3y from each side.
$x = 7 - 3y$	Simplify.

Find the value of y by substituting $7 - 3y$ for x in the second equation.

$2x - 4y = -6$	Second equation
$2(7 - 3y) - 4y = -6$	$x = 7 - 3y$
$14 - 6y - 4y = -6$	Distributive Property
$14 - 10y = -6$	Combine like terms.
$14 - 10y - 14 = -6 - 14$	Subtract 14 from each side.
$-10y = -20$	Simplify.
$y = 2$	Divide each side by -10 and simplify.

Use $y = 2$ to find the value of x.

$x = 7 - 3y$
$x = 7 - 3(2)$
$x = 1$

The solution is $(1, 2)$.

Exercises

Use substitution to solve each system of equations.

1. $y = 4x$
$3x - y = 1$

2. $x = 2y$
$y = x - 2$

3. $x = 2y - 3$
$x = 2y + 4$

4. $x - 2y = -1$
$3y = x + 4$

5. $x - 4y = 1$
$2x - 8y = 2$

6. $x + 2y = 0$
$3x + 4y = 4$

7. $2b = 6a - 14$
$3a - b = 7$

8. $x + y = 16$
$2y = -2x + 2$

9. $y = -x + 3$
$2y + 2x = 4$

10. $x = 2y$
$0.25x + 0.5y = 10$

11. $x - 2y = -5$
$x + 2y = -1$

12. $-0.2x + y = 0.5$
$0.4x + y = 1.1$

6-2　Study Guide and Intervention (continued)

Substitution

Solve Real-World Problems Substitution can also be used to solve real-world problems involving systems of equations. It may be helpful to use tables, charts, diagrams, or graphs to help you organize data.

Example CHEMISTRY **How much of a 10% saline solution should be mixed with a 20% saline solution to obtain 1000 milliliters of a 12% saline solution?**

Let s = the number of milliliters of 10% saline solution.
Let t = the number of milliliters of 20% saline solution.
Use a table to organize the information.

	10% saline	20% saline	12% saline
Total milliliters	s	t	1000
Milliliters of saline	$0.10s$	$0.20t$	$0.12(1000)$

Write a system of equations.
$s + t = 1000$
$0.10s + 0.20t = 0.12(1000)$
Use substitution to solve this system.

$s + t = 1000$	First equation
$s = 1000 - t$	Solve for s.
$0.10s + 0.20t = 0.12(1000)$	Second equation
$0.10(1000 - t) + 0.20t = 0.12(1000)$	$s = 1000 - t$
$100 - 0.10t + 0.20t = 0.12(1000)$	Distributive Property
$100 + 0.10t = 0.12(1000)$	Combine like terms.
$0.10t = 20$	Simplify.
$\dfrac{0.10t}{0.10} = \dfrac{20}{0.10}$	Divide each side by 0.10.
$t = 200$	Simplify.
$s + t = 1000$	First equation
$s + 200 = 1000$	$t = 200$
$s = 800$	Solve for s.

800 milliliters of 10% solution and 200 milliliters of 20% solution should be used.

Exercises

1. **SPORTS** At the end of the 2007–2008 football season, 38 Super Bowl games had been played with the current two football leagues, the American Football Conference (AFC) and the National Football Conference (NFC). The NFC won two more games than the AFC. How many games did each conference win?

2. **CHEMISTRY** A lab needs to make 100 gallons of an 18% acid solution by mixing a 12% acid solution with a 20% solution. How many gallons of each solution are needed?

3. **GEOMETRY** The perimeter of a triangle is 24 inches. The longest side is 4 inches longer than the shortest side, and the shortest side is three-fourths the length of the middle side. Find the length of each side of the triangle.

6-3 Study Guide and Intervention

Elimination Using Addition and Subtraction

Elimination Using Addition In systems of equations in which the coefficients of the x or y terms are additive inverses, solve the system by adding the equations. Because one of the variables is eliminated, this method is called **elimination**.

Example 1 Use elimination to solve the system of equations.
$$x - 3y = 7$$
$$3x + 3y = 9$$

Write the equations in column form and add to eliminate y.

$$\begin{array}{r} x - 3y = 7 \\ (+)\ 3x + 3y = 9 \\ \hline 4x\quad\ \ = 16 \end{array}$$

Solve for x.
$$\frac{4x}{4} = \frac{16}{4}$$
$$x = 4$$

Substitute 4 for x in either equation and solve for y.
$$4 - 3y = 7$$
$$4 - 3y - 4 = 7 - 4$$
$$-3y = 3$$
$$\frac{-3y}{-3} = \frac{3}{-3}$$
$$y = -1$$

The solution is $(4, -1)$.

Example 2 The sum of two numbers is 70 and their difference is 24. Find the numbers.

Let x represent one number and y represent the other number.

$$\begin{array}{r} x + y = 70 \\ (+)\ x - y = 24 \\ \hline 2x\quad\ \ = 94 \end{array}$$
$$\frac{2x}{2} = \frac{94}{2}$$
$$x = 47$$

Substitute 47 for x in either equation.
$$47 + y = 70$$
$$47 + y - 47 = 70 - 47$$
$$y = 23$$

The numbers are 47 and 23.

Exercises

Use elimination to solve each system of equations.

1. $x + y = -4$
$x - y = 2$

2. $2x - 3y = 14$
$x + 3y = -11$

3. $3x - y = -9$
$-3x - 2y = 0$

4. $-3x - 4y = -1$
$3x - y = -4$

5. $3x + y = 4$
$2x - y = 6$

6. $-2x + 2y = 9$
$2x - y = -6$

7. $2x + 2y = -2$
$3x - 2y = 12$

8. $4x - 2y = -1$
$-4x + 4y = -2$

9. $x - y = 2$
$x + y = -3$

10. $2x - 3y = 12$
$4x + 3y = 24$

11. $-0.2x + y = 0.5$
$0.2x + 2y = 1.6$

12. $0.1x + 0.3y = 0.9$
$0.1x - 0.3y = 0.2$

13. Rema is older than Ken. The difference of their ages is 12 and the sum of their ages is 50. Find the age of each.

14. The sum of the digits of a two-digit number is 12. The difference of the digits is 2. Find the number if the units digit is larger than the tens digit.

6-3   Study Guide and Intervention (continued)

Elimination Using Addition and Subtraction

Elimination Using Subtraction In systems of equations where the coefficients of the x or y terms are the same, solve the system by subtracting the equations.

Example   **Use elimination to solve the system of equations.**

$2x - 3y = 11$
$5x - 3y = 14$

$2x - 3y = 11$	Write the equations in column form and subtract.
$(-)\ 5x - 3y = 14$	
$\overline{\quad -3x \quad\quad = -3}$	Subtract the two equations. y is eliminated.
$\dfrac{-3x}{-3} = \dfrac{-3}{-3}$	Divide each side by -3.
$x = 1$	Simplify.
$2(1) - 3y = 11$	Substitute 1 for x in either equation.
$2 - 3y = 11$	Simplify.
$2 - 3y - 2 = 11 - 2$	Subtract 2 from each side.
$-3y = 9$	Simplify.
$\dfrac{-3y}{-3} = \dfrac{9}{-3}$	Divide each side by -3.
$y = -3$	Simplify.

The solution is $(1, -3)$.

Exercises

Use elimination to solve each system of equations.

1. $6x + 5y = 4$
$6x - 7y = -20$

2. $3m - 4n = -14$
$3m + 2n = -2$

3. $3a + b = 1$
$a + b = 3$

4. $-3x - 4y = -23$
$-3x + y = 2$

5. $x - 3y = 11$
$2x - 3y = 16$

6. $x - 2y = 6$
$x + y = 3$

7. $2a - 3b = -13$
$2a + 2b = 7$

8. $4x + 2y = 6$
$4x + 4y = 10$

9. $5x - y = 6$
$5x + 2y = 3$

10. $6x - 3y = 12$
$4x - 3y = 24$

11. $x + 2y = 3.5$
$x - 3y = -9$

12. $0.2x + y = 0.7$
$0.2x + 2y = 1.2$

13. The sum of two numbers is 70. One number is ten more than twice the other number. Find the numbers.

14. GEOMETRY Two angles are supplementary. The measure of one angle is $10°$ more than three times the other. Find the measure of each angle.

6-4 Study Guide and Intervention

Elimination Using Multiplication

Elimination Using Multiplication Some systems of equations cannot be solved simply by adding or subtracting the equations. In such cases, one or both equations must first be multiplied by a number before the system can be solved by elimination.

Example 1 Use elimination to solve the system of equations.

$x + 10y = 3$
$4x + 5y = 5$

If you multiply the second equation by -2, you can eliminate the y terms.

$$\begin{array}{r} x + 10y = 3 \\ (+)\ -8x - 10y = -10 \\ \hline -7x \qquad\quad = -7 \end{array}$$

$$\frac{-7x}{-7} = \frac{-7}{-7}$$
$$x = 1$$

Substitute 1 for x in either equation.

$$1 + 10y = 3$$
$$1 + 10y - 1 = 3 - 1$$
$$10y = 2$$
$$\frac{10y}{10} = \frac{2}{10}$$
$$y = \frac{1}{5}$$

The solution is $\left(1, \dfrac{1}{5}\right)$.

Example 2 Use elimination to solve the system of equations.

$3x - 2y = -7$
$2x - 5y = 10$

If you multiply the first equation by 2 and the second equation by -3, you can eliminate the x terms.

$$\begin{array}{r} 6x - 4y\ = -14 \\ (+)\ -6x + 15y\ = -30 \\ \hline 11y\ = -44 \end{array}$$

$$\frac{11y}{11} = -\frac{44}{11}$$
$$y = -4$$

Substitute -4 for y in either equation.

$$3x - 2(-4) = -7$$
$$3x + 8 = -7$$
$$3x + 8 - 8 = -7 - 8$$
$$3x = -15$$
$$\frac{3x}{3} = -\frac{15}{3}$$
$$x = -5$$

The solution is $(-5, -4)$.

Exercises

Use elimination to solve each system of equations.

1. $2x + 3y = 6$
   $x + 2y = 5$

2. $2m + 3n = 4$
   $-m + 2n = 5$

3. $3a - b = 2$
   $a + 2b = 3$

4. $4x + 5y = 6$
   $6x - 7y = -20$

5. $4x - 3y = 22$
   $2x - y = 10$

6. $3x - 4y = -4$
   $x + 3y = -10$

7. $4x - y = 9$
   $5x + 2y = 8$

8. $4a - 3b = -8$
   $2a + 2b = 3$

9. $2x + 2y = 5$
   $4x - 4y = 10$

10. $6x - 4y = -8$
   $4x + 2y = -3$

11. $4x + 2y = -5$
   $-2x - 4y = 1$

12. $2x + y = 3.5$
   $-x + 2y = 2.5$

13. GARDENING The length of Sally's garden is 4 meters greater than 3 times the width. The perimeter of her garden is 72 meters. What are the dimensions of Sally's garden?

6-4 Study Guide and Intervention (continued)

Elimination Using Multiplication

Solve Real-World Problems Sometimes it is necessary to use multiplication before elimination in real-world problems.

Example **CANOEING** During a canoeing trip, it takes Raymond 4 hours to paddle 12 miles upstream. It takes him 3 hours to make the return trip paddling downstream. Find the speed of the canoe in still water.

Read   You are asked to find the speed of the canoe in still water.

Solve   Let c = the rate of the canoe in still water.
Let w = the rate of the water current.

	r	t	d	$r \cdot t = d$
Against the Current	$c - w$	4	12	$(c - w)4 = 12$
With the Current	$c + w$	3	12	$(c + w)3 = 12$

So, our two equations are $4c - 4w = 12$ and $3c + 3w = 12$.

Use elimination with multiplication to solve the system. Since the problem asks for c, eliminate w.

$4c - 4w = 12 \Rightarrow$ Multiply by 3 $\Rightarrow$ $12c - 12w = 36$
$3c + 3w = 12 \Rightarrow$ Multiply by 4 $\Rightarrow$ $\underline{(+)\ 12c + 12w = 48}$
$ 24c = 84$ *w* is eliminated.

$$\frac{24c}{24} = \frac{84}{24}$$ Divide each side by 24.

$$c = 3.5$$ Simplify.

The rate of the canoe in still water is 3.5 miles per hour.

Exercises

1. **FLIGHT** An airplane traveling with the wind flies 450 miles in 2 hours. On the return trip, the plane takes 3 hours to travel the same distance. Find the speed of the airplane if the wind is still.

2. **FUNDRAISING** Benji and Joel are raising money for their class trip by selling gift wrapping paper. Benji raises $39 by selling 5 rolls of red wrapping paper and 2 rolls of foil wrapping paper. Joel raises $57 by selling 3 rolls of red wrapping paper and 6 rolls of foil wrapping paper. For how much are Benji and Joel selling each roll of red and foil wrapping paper?

6-5 Study Guide and Intervention

Applying Systems of Linear Equations

Determine The Best Method You have learned five methods for solving systems of linear equations: graphing, substitution, elimination using addition, elimination using subtraction, and elimination using multiplication. For an exact solution, an algebraic method is best.

Example At a baseball game, Henry bought 3 hotdogs and a bag of chips for $14. Scott bought 2 hotdogs and a bag of chips for $10. Each of the boys paid the same price for their hotdogs, and the same price for their chips. The following system of equations can be used to represent the situation. Determine the best method to solve the system of equations. Then solve the system.

$3x + y = 14$
$2x + y = 10$

• Since neither the coefficients of x nor the coefficients of y are additive inverses, you cannot use elimination using addition.

• Since the coefficient of y in both equations is 1, you can use elimination using subtraction. You could also use the substitution method or elimination using multiplication

The following solution uses elimination by subtraction to solve this system.

$$\begin{array}{ll} 3x + y = 14 & \text{Write the equations in column form and subtract.} \\ (-)\ 2x + (-)\ y = (-)10 & \\ \hline x \qquad = 4 & \text{The variable } y \text{ is eliminated.} \\ 3(4) + y = 14 & \text{Substitute the value for } x \text{ back into the first equation.} \\ y = 2 & \text{Solve for } y. \end{array}$$

This means that hot dogs cost $4 each and a bag of chips costs $2.

Exercises

Determine the best method to solve each system of equations. Then solve the system.

1. $5x + 3y = 16$
$3x - 5y = -4$

2. $3x - 5y = 7$
$2x + 5y = 13$

3. $y + 3x = 24$
$5x - y = 8$

4. $-11x - 10y = 17$
$5x - 7y = 50$

6-5 **Study Guide and Intervention** (continued)

Applying Systems of Linear Equations

Apply Systems Of Linear Equations When applying systems of linear equations to problem situations, it is important to analyze each solution in the context of the situation.

Example BUSINESS A T-shirt printing company sells T-shirts for $15 each. The company has a fixed cost for the machine used to print the T-shirts and an additional cost per T-shirt. Use the table to estimate the number of T-shirts the company must sell in order for the income to equal expenses.

T-shirt Printing Cost	
printing machine	$3000.00
blank T-shirt	$5.00

Understand You know the initial income and the initial expense and the rates of change of each quantity with each T-shirt sold.

Plan Write an equation to represent the income and the expenses. Then solve to find how many T-shirts need to be sold for both values to be equal.

Solve Let x = the number of T-shirts sold and let y = the total amount.

	total amount	initial amount	rate of change times number of T-shirts sold
income	y =	0 +	$15x$
expenses	y =	3000 +	$5x$

You can use substitution to solve this system.

$y = 15x$ The first equation.

$15x = 3000 + 5x$ Substitute the value for y into the second equation.

$10x = 3000$ Subtract 10x from each side and simplify.

$x = 300$ Divide each side by 10 and simplify.

This means that if 300 T-shirts are sold, the income and expenses of the T-shirt company are equal.

Check Does this solution make sense in the context of the problem? After selling 300 T-shirts, the income would be about 300 × $15 or $4500. The costs would be about $3000 + 300 × $5 or $4500.

Exercises

Refer to the example above. If the costs of the T-shirt company change to the given values and the selling price remains the same, determine the number of T-shirts the company must sell in order for income to equal expenses.

1. printing machine: $5000.00; T-shirt: $10.00 each

2. printing machine: $2100.00; T-shirt: $8.00 each

3. printing machine: $8800.00; T-shirt: $4.00 each

4. printing machine: $1200.00; T-shirt: $12.00 each

6-6 Study Guide and Intervention

Systems of Inequalities

Systems of Inequalities The solution of a **system of inequalities** is the set of all ordered pairs that satisfy both inequalities. If you graph the inequalities in the same coordinate plane, the solution is the region where the graphs overlap.

Example 1 Solve the system of inequalities by graphing.

$y > x + 2$
$y \leq -2x - 1$

The solution includes the ordered pairs in the intersection of the graphs. This region is shaded at the right. The graphs of $y = x + 2$ and $y = -2x - 1$ are boundaries of this region. The graph of $y = x + 2$ is dashed and is not included in the graph of $y > x + 2$.

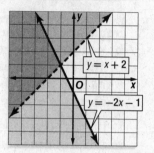

Example 2 Solve the system of inequalities by graphing.

$x + y > 4$
$x + y < -1$

The graphs of $x + y = 4$ and $x + y = -1$ are parallel. Because the two regions have no points in common, the system of inequalities has no solution.

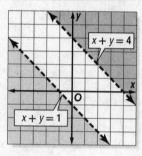

Exercises

Solve each system of inequalities by graphing.

1. $y > -1$
$x < 0$

2. $y > -2x + 2$
$y \leq x + 1$

3. $y < x + 1$
$3x + 4y \geq 12$

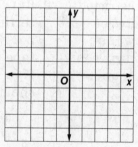

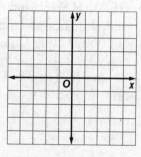

4. $2x + y \geq 1$
$x - y \geq -2$

5. $y \leq 2x + 3$
$y \geq -1 + 2x$

6. $5x - 2y < 6$
$y > -x + 1$

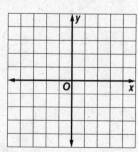

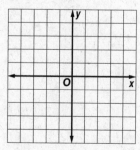

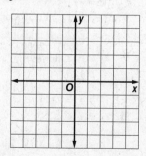

6-6 Study Guide and Intervention (continued)

Systems of Inequalities

Apply Systems of Inequalities In real-world problems, sometimes only whole numbers make sense for the solution, and often only positive values of x and y make sense.

Example **BUSINESS** AAA Gem Company produces necklaces and bracelets. In a 40-hour week, the company has 400 gems to use. A necklace requires 40 gems and a bracelet requires 10 gems. It takes 2 hours to produce a necklace and a bracelet requires one hour. How many of each type can be produced in a week?

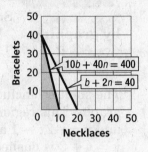

Let n = the number of necklaces that will be produced and b = the number of bracelets that will be produced. Neither n or b can be a negative number, so the following system of inequalities represents the conditions of the problems.

$n \geq 0$

$b \geq 0$

$b + 2n \leq 40$

$10b + 40n \leq 400$

The solution is the set ordered pairs in the intersection of the graphs. This region is shaded at the right. Only whole-number solutions, such as (5, 20) make sense in this problem.

Exercises

For each exercise, graph the solution set. List three possible solutions to the problem.

1. **HEALTH** Mr. Flowers is on a restricted diet that allows him to have between 1600 and 2000 Calories per day. His daily fat intake is restricted to between 45 and 55 grams. What daily Calorie and fat intakes are acceptable?

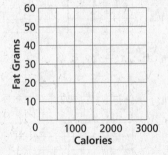

2. **RECREATION** Maria had $150 in gift certificates to use at a record store. She bought fewer than 20 recordings. Each tape cost $5.95 and each CD cost $8.95. How many of each type of recording might she have bought?

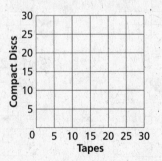

7-1 Study Guide and Intervention

Multiplication Properties of Exponents

Multiply Monomials A **monomial** is a number, a variable, or the product of a number and one or more variables with nonnegative integer exponents. An expression of the form x^n is called a **power** and represents the product you obtain when x is used as a factor n times. To multiply two powers that have the same base, add the exponents.

Product of Powers	For any number a and all integers m and n, $a^m \cdot a^n = a^{m+n}$.

Example 1 Simplify $(3x^6)(5x^2)$.

$(3x^6)(5x^2) = (3)(5)(x^6 \cdot x^2)$ Group the coefficients

                             and the variables

$\qquad\qquad = (3 \cdot 5)(x^{6+2})$ Product of Powers

$\qquad\qquad = 15x^8$ Simplify.

The product is $15x^8$.

Example 2 Simplify $(-4a^3b)(3a^2b^5)$.

$(-4a^3b)(3a^2b^5) = (-4)(3)(a^3 \cdot a^2)(b \cdot b^5)$

$\qquad\qquad\qquad = -12(a^{3+2})(b^{1+5})$

$\qquad\qquad\qquad = -12a^5b^6$

The product is $-12a^5b^6$.

Exercises

Simplify each expression.

1. $y(y^5)$

2. $n^2 \cdot n^7$

3. $(-7x^2)(x^4)$

4. $x(x^2)(x^4)$

5. $m \cdot m^5$

6. $(-x^3)(-x^4)$

7. $(2a^2)(8a)$

8. $(rn)(rn^3)(n^2)$

9. $(x^2y)(4xy^3)$

10. $\frac{1}{3}(2a^3b)(6b^3)$

11. $(-4x^3)(-5x^7)$

12. $(-3j^2k^4)(2jk^6)$

13. $(5a^2bc^3)\left(\frac{1}{5}abc^4\right)$

14. $(-5xy)(4x^2)(y^4)$

15. $(10x^3yz^2)(-2xy^5z)$

7-1   Study Guide and Intervention (continued)

Multiplication Properties of Exponents

Simplify Expressions An expression of the form $(x^m)^n$ is called a **power of a power** and represents the product you obtain when x^m is used as a factor n times. To find the power of a power, multiply exponents.

Power of a Power	For any number a and any integers m and p, $(a^m)^p = a^{mp}$.
Power of a Product	For any numbers a and b and any integer m, $(ab)^m = a^m b^m$.

We can combine and use these properties to simplify expressions involving monomials.

Example   Simplify $(-2ab^2)^3(a^2)^4$.

$$
\begin{aligned}
(-2ab^2)^3(a^2)^4 &= (-2ab^2)^3(a^8) && \text{Power of a Power} \\
&= (-2)^3(a^3)(b^2)^3(a^8) && \text{Power of a Product} \\
&= (-2)^3(a^3)(a^8)(b^2)^3 && \text{Group the coefficients and the variables} \\
&= (-2)^3(a^{11})(b^2)^3 && \text{Product of Powers} \\
&= -8a^{11}b^6 && \text{Power of a Power}
\end{aligned}
$$

The product is $-8a^{11}b^6$.

Exercises

Simplify each expression.

1. $(y^5)^2$

2. $(n^7)^4$

3. $(x^2)^5(x^3)$

4. $-3(ab^4)^3$

5. $(-3ab^4)^3$

6. $(4x^2b)^3$

7. $(4a^2)^2(b^3)$

8. $(4x)^2(b^3)$

9. $(x^2y^4)^5$

10. $(2a^3b^2)(b^3)^2$

11. $(-4xy)^3(-2x^2)^3$

12. $(-3j^2k^3)^2(2j^2k)^3$

13. $(25a^2b)^3\left(\dfrac{1}{5}\,abf\right)^2$

14. $(2xy)^2(-3x^2)(4y^4)$

15. $(2x^3y^2z^2)^3(x^2z)^4$

16. $(-2n^6y^5)(-6n^3y^2)(ny)^3$

17. $(-3a^3n^4)(-3a^3n)^4$

18. $-3(2x)^4(4x^5y)^2$

7-2 Study Guide and Intervention

Division Properties of Exponents

Divide Monomials To divide two powers with the same base, subtract the exponents.

Quotient of Powers	For all integers m and n and any nonzero number a, $\dfrac{a^m}{a^n} = a^{m-n}$.
Power of a Quotient	For any integer m and any real numbers a and b, $b \neq 0$, $\left(\dfrac{a}{b}\right)^m = \dfrac{a^m}{b^m}$.

Example 1 Simplify $\dfrac{a^4 b^7}{ab^2}$. Assume that no denominator equals zero.

$\dfrac{a^4 b^7}{ab^2} = \left(\dfrac{a^4}{a}\right)\left(\dfrac{b^7}{b^2}\right)$ Group powers with the same base.

$= (a^{4-1})(b^{7-2})$ Quotient of Powers

$= a^3 b^5$ Simplify.

The quotient is $a^3 b^5$.

Example 2 Simplify $\left(\dfrac{2a^3 b^5}{3b^2}\right)^3$. Assume that no denominator equals zero.

$\left(\dfrac{2a^3 b^5}{3b^2}\right)^3 = \dfrac{(2a^3 b^5)^3}{(3b^2)^3}$ Power of a Quotient

$= \dfrac{2^3 (a^3)^3 (b^5)^3}{(3)^3 (b^2)^3}$ Power of a Product

$= \dfrac{8a^9 b^{15}}{27b^6}$ Power of a Power

$= \dfrac{8a^9 b^9}{27}$ Quotient of Powers

The quotient is $\dfrac{8a^9 b^9}{27}$.

Exercises

Simplify each expression. Assume that no denominator equals zero.

1. $\dfrac{5^5}{5^2}$

2. $\dfrac{m^6}{m^4}$

3. $\dfrac{p^5 n^4}{p^2 n}$

4. $\dfrac{a^2}{a}$

5. $\dfrac{x^5 y^3}{x^5 y^2}$

6. $\dfrac{-2y^7}{14y^5}$

7. $\dfrac{xy^6}{y^4 x}$

8. $\left(\dfrac{2a^2 b}{a}\right)^3$

9. $\left(\dfrac{4p^4 r^4}{3p^2 r^2}\right)^3$

10. $\left(\dfrac{2r^5 w^3}{r^4 w^3}\right)^4$

11. $\left(\dfrac{3r^6 n^3}{2r^5 n}\right)^4$

12. $\dfrac{r^7 n^7 t^2}{n^3 r^3 t^2}$

7-2 Study Guide and Intervention *(continued)*

Division Properties of Exponents

Negative Exponents Any nonzero number raised to the zero power is 1; for example, $(-0.5)^0 = 1$. Any nonzero number raised to a negative power is equal to the reciprocal of the number raised to the opposite power; for example, $6^{-3} = \dfrac{1}{6^3}$. These definitions can be used to simplify expressions that have negative exponents.

Zero Exponent	For any nonzero number a, $a^0 = 1$.
Negative Exponent Property	For any nonzero number a and any integer n, $a^{-n} = \dfrac{1}{a^n}$ and $\dfrac{1}{a^{-n}} = a^n$.

The simplified form of an expression containing negative exponents must contain only positive exponents.

Example Simplify $\dfrac{4a^{-3}b^6}{16a^2b^6c^{-5}}$. **Assume that no denominator equals zero.**

$\dfrac{4a^{-3}b^6}{16a^2b^6c^{-5}} = \left(\dfrac{4}{16}\right)\left(\dfrac{a^{-3}}{a^2}\right)\left(\dfrac{b^6}{b^6}\right)\left(\dfrac{1}{c^{-5}}\right)$ Group powers with the same base.

$\qquad = \dfrac{1}{4}\,(a^{-3-2})(b^{6-6})(c^5)$ Quotient of Powers and Negative Exponent Properties

$\qquad = \dfrac{1}{4}\,a^{-5}b^0c^5$ Simplify.

$\qquad = \dfrac{1}{4}\left(\dfrac{1}{a^5}\right)(1)c^5$ Negative Exponent and Zero Exponent Properties

$\qquad = \dfrac{c^5}{4a^5}$ Simplify.

The solution is $\dfrac{c^5}{4a^5}$.

Exercises

Simplify each expression. Assume that no denominator equals zero.

1. $\dfrac{2^2}{2^{-3}}$

2. $\dfrac{m}{m^{-4}}$

3. $\dfrac{p^{-8}}{p^3}$

4. $\dfrac{b^{-4}}{b^{-5}}$

5. $\dfrac{(-x^{-1}y)^0}{4w^{-1}y^2}$

6. $\dfrac{(a^2b^3)^2}{(ab)^{-2}}$

7. $\dfrac{x^4y^0}{x^{-2}}$

8. $\dfrac{(6a^{-1}b)^2}{(b^2)^4}$

9. $\dfrac{(3rt)^2u^{-4}}{r^{-1}t^2u^7}$

10. $\dfrac{m^{-3}t^{-5}}{(m^2t^3)^{-1}}$

11. $\left(\dfrac{4m^2n^2}{8m^{-1}\ell}\right)^0$

12. $\dfrac{(-2mn^2)^{-3}}{4m^{-6}n^4}$

7-3 Study Guide and Intervention

Rational Exponents

Rational Exponents For any real numbers a and b and any positive integer n, if $a^n = b$, then a is an nth root of b. Rational exponents can be used to represent nth roots.

Square Root	$b^{\frac{1}{2}} = \sqrt{b}$
Cube Root	$b^{\frac{1}{3}} = \sqrt[3]{b}$
nth Root	$b^{\frac{1}{n}} = \sqrt[n]{b}$

Example 1 Write $(6xy)^{\frac{1}{2}}$ in radical form.

$(6xy)^{\frac{1}{2}} = \sqrt{6xy}$ Definition of $b^{\frac{1}{2}}$

Example 2 Simplify $625^{\frac{1}{4}}$.

$625^{\frac{1}{4}} = \sqrt[4]{625}$ $b^{\frac{1}{n}} = \sqrt[n]{b}$

$= \sqrt[4]{5 \cdot 5 \cdot 5 \cdot 5}$ $625 = 5^4$

$= 5$ Simplify.

Exercises

Write each expression in radical form, or write each radical in exponential form.

1. $14^{\frac{1}{2}}$

2. $5x^{\frac{1}{2}}$

3. $17y^{\frac{1}{2}}$

4. $12^{\frac{1}{2}}$

5. $19ab^{\frac{1}{2}}$

6. $\sqrt{17}$

7. $\sqrt{12n}$

8. $\sqrt{18b}$

9. $\sqrt{37}$

Simplify.

10. $\sqrt[3]{343}$

11. $\sqrt[5]{1024}$

12. $512^{\frac{1}{3}}$

13. $\sqrt[4]{2401}$

14. $\sqrt[6]{64}$

15. $243^{\frac{1}{5}}$

16. $\sqrt[3]{1331}$

17. $\sqrt[4]{6561}$

18. $4096^{\frac{1}{4}}$

7-3 Study Guide and Intervention *(continued)*

Rational Exponents

Solve Exponential Equations In an **exponential equation**, variables occur as exponents. Use the Power Property of Equality and the other properties of exponents to solve exponential equations.

Example **Solve $1024^{x-1} = 4$.**

$1024^{x-1} = 4$	Original equation
$(4^5)^{x-1} = 4$	Rewrite 1024 as 4^5.
$4^{5x-5} = 4^1$	Power of a Power, Distributive Property
$5x - 5 = 1$	Power Property of Equality
$5x = 6$	Add 5 to each side.
$x = \dfrac{6}{5}$	Divide each side by 5.

Exercises

Solve each equation.

1. $2^x = 128$

2. $3^{3x+1} = 81$

3. $4^{x-3} = 32$

4. $5^x = 15{,}625$

5. $6^{3x+2} = 216$

6. $4^{5x-3} = 16$

7. $8^x = 4096$

8. $9^{3x+3} = 6561$

9. $11^{x-1} = 1331$

10. $3^x = 6561$

11. $2^{5x+4} = 512$

12. $7^{x-2} = 343$

13. $8^x = 262{,}144$

14. $5^{5x} = 3125$

15. $9^{2x-6} = 6561$

16. $7^x = 2401$

17. $7^{3x} = 117{,}649$

18. $6^{2x-7} = 7776$

19. $9^x = 729$

20. $8^{3x+1} = 4096$

21. $13^{3x-8} = 28{,}561$

7-4　Study Guide and Intervention

Scientific Notation

Scientific Notation Very large and very small numbers are often best represented using a method known as **scientific notation**. Numbers written in scientific notation take the form $a \times 10^n$, where $1 \le a < 10$ and n is an integer. Any number can be written in scientific notation.

Example 1 **Express 34,020,000,000 in scientific notation.**

Step 1 Move the decimal point until it is to the right of the first nonzero digit. The result is a real number a. Here, $a = 3.402$.

Step 2 Note the number of places n and the direction that you moved the decimal point. The decimal point moved 10 places to the left, so $n = 10$.

Step 3 Because the decimal moved to the left, write the number as $a \times 10^n$.

$34{,}020{,}000{,}000 = 3.4020000000 \times 10^{10}$

Step 4 Remove the extra zeros. 3.402×10^{10}

Example 2 **Express 4.11×10^{-6} in standard notation.**

Step 1 The exponent is -6, so $n = -6$.

Step 2 Because $n < 0$, move the decimal point 6 places to the left.

$4.11 \times 10^{-6} \Rightarrow .00000411$

Step 3 $4.11 \times 10^{-6} \Rightarrow 0.00000411$

Rewrite; insert a 0 before the decimal point.

Exercises

Express each number in scientific notation.

1. 5,100,000

2. 80,300,000,000

3. 14,250,000

4. 68,070,000,000,000

5. 14,000

6. 901,050,000,000

7. 0.0049

8. 0.000301

9. 0.0000000519

10. 0.000000185

11. 0.002002

12. 0.00000771

Express each number in standard form.

13. 4.91×10^4

14. 3.2×10^{-5}

15. 6.03×10^8

16. 2.001×10^{-6}

17. 1.00024×10^{10}

18. 5×10^5

19. 9.09×10^{-5}

20. 3.5×10^{-2}

21. 1.7087×10^7

7-4 Study Guide and Intervention (continued)

Scientific Notation

Products and Quotients in Scientific Notation You can use scientific notation to simplify multiplying and dividing very large and very small numbers.

Example 1 Evaluate $(9.2 \times 10^{-3}) \times (4 \times 10^8)$. Express the result in both scientific notation and standard form.

$(9.2 \times 10^{-3})(4 \times 10^8)$	Original expression
$= (9.2 \times 4)(10^{-3} \times 10^8)$	Commutative and Associative Properties
$= 36.8 \times 10^5$	Product of Powers
$= (3.68 \times 10^1) \times 10^5$	$36.8 = 3.68 \times 10$
$= 3.68 \times 10^6$	Product of Powers
$= 3,680,000$	Standard Form

Example 2 Evaluate $\dfrac{(2.76 \times 10^7)}{(6.9 \times 10^5)}$. Express the result in both scientific notation and standard form.

$\dfrac{(2.76 \times 10^7)}{(6.9 \times 10^5)} = \left(\dfrac{2.76}{6.9}\right)\left(\dfrac{10^7}{10^5}\right)$	Product rule for fractions
$= 0.4 \times 10^2$	Quotient of Powers
$= 4.0 \times 10^{-1} \times 10^2$	$0.4 = 4.0 \times 10^{-1}$
$= 4.0 \times 10^1$	Product of Powers
$= 40$	Standard form

Exercises

Evaluate each product. Express the results in both scientific notation and standard form.

1. $(3.4 \times 10^3)(5 \times 10^4)$

2. $(2.8 \times 10^{-4})(1.9 \times 10^7)$

3. $(6.7 \times 10^{-7})(3 \times 10^3)$

4. $(8.1 \times 10^5)(2.3 \times 10^{-3})$

5. $(1.2 \times 10^{-4})^2$

6. $(5.9 \times 10^5)^2$

Evaluate each quotient. Express the results in both scientific notation and standard form.

7. $\dfrac{(4.9 \times 10^{-3})}{(2.5 \times 10^{-4})}$

8. $\dfrac{5.8 \times 10^4}{5 \times 10^{-2}}$

9. $\dfrac{(1.6 \times 10^5)}{(4 \times 10^{-4})}$

10. $\dfrac{8.6 \times 10^6}{1.6 \times 10^{-3}}$

11. $\dfrac{(4.2 \times 10^{-2})}{(6 \times 10^{-7})}$

12. $\dfrac{8.1 \times 10^5}{2.7 \times 10^4}$

7-5 Study Guide and Intervention

Exponential Functions

Graph Exponential Functions

Exponential Function	a function defined by an equation of the form $y = ab^x$, where $a \neq 0$, $b > 0$, and $b \neq 1$

You can use values of x to find ordered pairs that satisfy an exponential function. Then you can use the ordered pairs to graph the function.

Example 1 Graph $y = 3^x$. Find the y-intercept and state the domain and range.

x	y
−2	$\frac{1}{9}$
−1	$\frac{1}{3}$
0	1
1	3
2	9

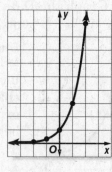

The y-intercept is 1.

The domain is all real numbers, and the range is all positive numbers.

Example 2 Graph $y = \left(\frac{1}{4}\right)^x$. Find the y-intercept and state the domain and range.

x	y
−2	16
−1	4
0	1
1	$\frac{1}{4}$
2	$\frac{1}{16}$

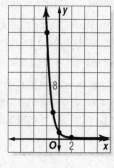

The y-intercept is 1.

The domain is all real numbers, and the range is all positive numbers.

Exercises

Graph each function. Find the y-intercept and state the domain and range.

1. $y = 0.3^x$

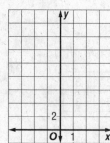

2. $y = 3x + 1$

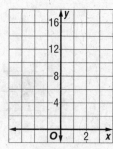

3. $y = \left(\frac{1}{3}\right)^x + 1$

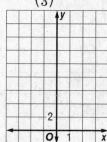

7-5 Study Guide and Intervention (continued)

Exponential Functions

Identify Exponential Behavior It is sometimes useful to know if a set of data is exponential. One way to tell is to observe the shape of the graph. Another way is to observe the pattern in the set of data.

Example Determine whether the set of data shown below displays exponential behavior. Write *yes* or *no*. Explain why or why not.

x	0	2	4	6	8	10
y	64	32	16	8	4	2

Method 1: Look for a Pattern

The domain values increase by regular intervals of 2, while the range values have a common factor of $\frac{1}{2}$. Since the domain values increase by regular intervals and the range values have a common factor, the data are probably exponential.

Method 2: Graph the Data

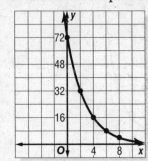

The graph shows rapidly decreasing values of y as x increases. This is characteristic of exponential behavior.

Exercises

Determine whether the set of data shown below displays exponential behavior. Write *yes* or *no*. Explain why or why not.

1.

x	0	1	2	3
y	5	10	15	20

2.

x	0	1	2	3
y	3	9	27	81

3.

x	−1	1	3	5
y	32	16	8	4

4.

x	−1	0	1	2	3
y	3	3	3	3	3

5.

x	−5	0	5	10
y	1	0.5	0.25	0.125

6.

x	0	1	2	3	4
y	$\frac{1}{3}$	$\frac{1}{9}$	$\frac{1}{27}$	$\frac{1}{81}$	$\frac{1}{243}$

7-6 Study Guide and Intervention
Growth and Decay

Exponential Growth Population increases and growth of monetary investments are examples of **exponential growth**. This means that an initial amount increases at a steady rate over time.

Exponential Growth	The general equation for exponential growth is $y = a(1 + r)^t$. • y represents the final amount. • a represents the initial amount. • r represents the rate of change expressed as a decimal. • t represents time.

Example 1 **POPULATION** The population of Johnson City in 2005 was 25,000. Since then, the population has grown at an average rate of 3.2% each year.

a. Write an equation to represent the population of Johnson City since 2005.

The rate 3.2% can be written as 0.032.

$y = a(1 + r)^t$

$y = 25{,}000(1 + 0.032)^t$

$y = 25{,}000(1.032)^t$

b. According to the equation, what will the population of Johnson City be in 2015?

In 2015 t will equal 2015 − 2005 or 10. Substitute 10 for t in the equation from part **a**.

$y = 25{,}000(1.032)^{10}$ $\qquad t = 10$

$\approx 34{,}256$

In 2015 the population of Johnson City will be about 34,256.

Example 2 **INVESTMENT** The Garcias have $12,000 in a savings account. The bank pays 3.5% interest on savings accounts, compounded monthly. Find the balance in 3 years.

The rate 3.5% can be written as 0.035. The special equation for compound interest is $A = P\left(1 + \frac{r}{n}\right)^{nt}$, where A represents the balance, P is the initial amount, r represents the annual rate expressed as a decimal, n represents the number of times the interest is compounded each year, and t represents the number of years the money is invested.

$A = P\left(1 + \frac{r}{n}\right)^{nt}$

$\quad = 12{,}000\left(1 + \frac{0.035}{12}\right)^{3(12)}$

$\quad \approx 13{,}326.49$

In three years, the balance of the account will be $13,326.49.

Exercises

1. **POPULATION** The population of the United States has been increasing at an average annual rate of 0.91%. If the population was about 303,146,000 in 2008, predict the population in 2012.

2. **INVESTMENT** Determine the value of an investment of $2500 if it is invested at an interest rate of 5.25% compounded monthly for 4 years.

3. **POPULATION** It is estimated that the population of the world is increasing at an average annual rate of 1.3%. If the 2008 population was about 6,641,000,000, predict the 2015 population.

4. **INVESTMENT** Determine the value of an investment of $100,000 if it is invested at an interest rate of 5.2% compounded quarterly for 12 years.

7-6　Study Guide and Intervention *(continued)*

Growth and Decay

Exponential Decay Radioactive decay and depreciation are examples of **exponential decay**. This means that an initial amount decreases at a steady rate over a period of time.

Exponential Decay	The general equation for exponential decay is $y = a(1 - r)^t$. • y represents the final amount. • a represents the initial amount. • r represents the rate of decay expressed as a decimal. • t represents time.

Example　**DEPRECIATION** The original price of a tractor was $45,000. The value of the tractor decreases at a steady rate of 12% per year.

a. Write an equation to represent the value of the tractor since it was purchased.

The rate 12% can be written as 0.12.

$y = a(1 - r)^t$　　　　General equation for exponential decay

$y = 45,000(1 - 0.12)^t$　　$a = 45,000$ and $r = 0.12$

$y = 45,000(0.88)^t$　　Simplify.

b. What is the value of the tractor in 5 years?

$y = 45,000(0.88)^t$　　　Equation for decay from part a

$y = 45,000(0.88)^5$　　　$t = 5$

$y \approx 23,747.94$　　　Use a calculator.

In 5 years, the tractor will be worth about $23,747.94.

Exercises

1. **POPULATION** The population of Bulgaria has been decreasing at an annual rate of 0.89%. If the population of Bulgaria was about 7,450,349 in the year 2005, predict its population in the year 2015.

2. **DEPRECIATION** Mr. Gossell is a machinist. He bought some new machinery for about $125,000. He wants to calculate the value of the machinery over the next 10 years for tax purposes. If the machinery depreciates at the rate of 15% per year, what is the value of the machinery (to the nearest $100) at the end of 10 years?

3. **ARCHAEOLOGY** The *half-life* of a radioactive element is defined as the time that it takes for one-half a quantity of the element to decay. Radioactive carbon-14 is found in all living organisms and has a half-life of 5730 years. Consider a living organism with an original concentration of carbon-14 of 100 grams.

   a. If the organism lived 5730 years ago, what is the concentration of carbon-14 today?

   b. If the organism lived 11,460 years ago, determine the concentration of carbon-14 today.

4. **DEPRECIATION** A new car costs $32,000. It is expected to depreciate 12% each year for 4 years and then depreciate 8% each year thereafter. Find the value of the car in 6 years.

7-7 Study Guide and Intervention
Geometric Sequences as Exponential Functions

Recognize Geometric Sequences A geometric sequence is a sequence in which each term after the first is found by multiplying the previous term by a nonzero constant r called the **common ratio**. The common ratio can be found by dividing any term by its previous term.

Example 1 Determine whether the sequence is *arithmetic, geometric,* or *neither: 21, 63, 189, 567, . . .*

Find the ratios of the consecutive terms. If the ratios are constant, the sequence is geometric.

21   63   189   567

$$\frac{63}{21} = \frac{189}{63} = \frac{567}{189} = 3$$

Because the ratios are constant, the sequence is geometric. The common ratio is 3.

Example 2 Find the next three terms in this geometric sequence: $-1215, 405, -135, 45, . . .$

Step 1 Find the common ratio.

-1215 405 -135 45

$$\frac{405}{-1215} = \frac{-135}{405} = \frac{45}{-135} = \frac{-1}{3}$$

The value of r is $-\frac{1}{3}$.

Step 2 Multiply each term by the common ratio to find the next three terms.

45 -15 5 $-\frac{5}{3}$

$\times\left(-\frac{1}{3}\right)$ $\times\left(-\frac{1}{3}\right)$ $\times\left(-\frac{1}{3}\right)$

The next three terms of the sequence are -15, 5, and $-\frac{5}{3}$.

Exercises

Determine whether each sequence is *arithmetic, geometric,* or *neither.* Explain.

1. 1, 2, 4, 8, . . .

2. 9, 14, 6, 11, . . .

3. $\frac{2}{3}, \frac{1}{3}, \frac{1}{6}, \frac{1}{12}, . . .$

4. $-2, 5, 12, 19, . . .$

Find the next three terms in each geometric sequence.

5. 648, -216, 72, . . .

6. 25, -5, 1, . . .

7. $\frac{1}{16}, \frac{1}{2}, 4, . . .$

8. 72, 36, 18, . . .

7-7 Study Guide and Intervention *(continued)*

Geometric Sequences as Exponential Functions

Geometric Sequences and Functions The nth term a_n of a geometric sequence with first term a_1 and common ratio r is given by the following formula, where n is any positive integer: $a_n = a_1 \cdot r^{n-1}$.

Example **a. Write an equation for the nth term of the geometric sequence 5, 20, 80, 320, . . .**

The first term of the sequence is 320. So, $a_1 = 320$. Now find the common ratio.

5   20   80   320

$\dfrac{20}{5} = \dfrac{80}{20} = \dfrac{320}{80} = 4$

The common ratio is 4. So, $r = 4$.

$a_n = a_1 \cdot r^{n-1}$ Formula for nth term

$a_n = 5 \cdot 4^{n-1}$ $a_1 = 5$ and $r = 4$

b. Find the seventh term of this sequence.

Because we are looking for the seventh term, $n = 7$.

$a_n = a_1 \cdot r^{n-1}$ Formula for nth term

$a_7 = 5 \cdot 4^{7-1}$ $n = 7$

$\quad = 5 \cdot 4^6$ Simplify.

$\quad = 5 \cdot 4096$ $4^6 = 4096$

$\quad = 20{,}480$ Multiply.

The seventh term of the sequence is 20,480.

Exercises

1. Write an equation for the nth term of the geometric sequence –2, 10, –50, ...
   Find the eleventh term of this sequence.

2. Write an equation for the nth term of the geometric sequence 512, 128, 32, ...
   Find the sixth term of this sequence.

3. Write an equation for the nth term of the geometric sequence $\dfrac{4}{9}$, 4, 36, ...
   Find the eighth term of this sequence.

4. Write an equation for the nth term of the geometric sequence 6, –54, 486, ...
   Find the ninth term of this sequence.

5. Write an equation for the nth term of the geometric sequence 100, 80, 64, ...
   Find the seventh term of this sequence.

6. Write an equation for the nth term of the geometric sequence $\dfrac{2}{5}, \dfrac{1}{10}, \dfrac{1}{40}, \dots$.
   Find the sixth term of this sequence.

7. Write an equation for the nth term of the geometric sequence $\dfrac{3}{8}, -\dfrac{3}{2}, 6, \dots$.
   Find the tenth term of this sequence.

8. Write an equation for the nth term of the geometric sequence –3, –21, –147, ...
   Find the fifth term of this sequence.

7-8 Study Guide and Intervention

Recursive Formulas

Using Recursive Formulas A **recursive formula** allows you to find the nth term of a sequence by performing operations on one or more of the terms that precede it.

Example Find the first five terms of the sequence in which $a_1 = 5$ and $a_n = -2a_{n-1} + 14$, if $n \geq 2$.

The given first term is $a_1 = 5$. Use this term and the recursive formula to find the next four terms.

$$a_2 = -2a_{2-1} + 14 \qquad n = 2 \qquad\qquad a_4 = -2a_{4-1} + 14 \qquad n = 4$$
$$= -2a_1 + 14 \qquad \text{Simplify.} \qquad\qquad = -2a_3 + 14 \qquad \text{Simplify.}$$
$$= -2(5) + 14 \text{ or } 4 \qquad a_1 = 5 \qquad\qquad = -2(6) + 14 \text{ or } 2 \qquad a_3 = 6$$

$$a_3 = -2a_{3-1} + 14 \qquad n = 3 \qquad\qquad a_5 = -2a_{5-1} + 14 \qquad n = 5$$
$$= -2a_2 + 14 \qquad \text{Simplify.} \qquad\qquad = -2a_4 + 14 \qquad \text{Simplify.}$$
$$= -2(4) + 14 \text{ or } 6 \qquad a_2 = 4 \qquad\qquad = -2(2) + 14 \text{ or } 10 \qquad a_4 = 2$$

The first five terms are 5, 4, 6, 2, and 10.

Exercises

Find the first five terms of each sequence.

1. $a_1 = -4, a_n = 3a_{n-1}, n \geq 2$ **2.** $a_1 = 5, a_n = 2a_{n-1}, n \geq 2$

3. $a_1 = 8, a_n = a_{n-1} - 6, n \geq 2$ **4.** $a_1 = -32, a_n = a_{n-1} + 13, n \geq 2$

5. $a_1 = 6, a_n = -3a_{n-1} + 20, n \geq 2$ **6.** $a_1 = -9, a_n = 2a_{n-1} + 11, n \geq 2$

7. $a_1 = 12, a_n = 2a_{n-1} - 10, n \geq 2$ **8.** $a_1 = -1, a_n = 4a_{n-1} + 3, n \geq 2$

9. $a_1 = 64, a_n = 0.5a_{n-1} + 8, n \geq 2$ **10.** $a_1 = 8, a_n = 1.5a_{n-1}, n \geq 2$

11. $a_1 = 400, a_n = \frac{1}{2}a_{n-1}, n \geq 2$ **12.** $a_1 = \frac{1}{4}, a_n = a_{n-1} + \frac{3}{4}, n \geq 2$

7-8   Study Guide and Intervention (continued)

Recursive Formulas

Writing Recursive Formulas Complete the following steps to write a recursive formula for an arithmetic or geometric sequence.

Step 1	Determine if the sequence is arithmetic or geometric by finding a common difference or a common ratio.
Step 2	Write a recursive formula. **Arithmetic Sequences** $a_n = a_{n-1} + d$, where d is the common difference **Geometric Sequences** $a_n = r \cdot a_{n-1}$, where r is the common ratio
Step 3	State the first term and the domain for n.

Example   Write a recursive formula for 216, 36, 6, 1, ...

Step 1   First subtract each term from the term that follows it.

$$216 - 36 = 180 \qquad 36 - 6 = 30 \qquad 6 - 1 = 5$$

There is no common difference. Check for a common ratio by dividing each term by the term that precedes it.

$$\frac{36}{216} = \frac{1}{6} \qquad\qquad \frac{6}{36} = \frac{1}{6} \qquad\qquad \frac{1}{6} = \frac{1}{6}$$

There is a common ratio of $\frac{1}{6}$. The sequence is geometric.

Step 2   Use the formula for a geometric sequence.

$a_n = r \cdot a_{n-1}$ Recursive formula for geometric sequence

$a_n = \frac{1}{6} a_{n-1}$ $r = \frac{1}{6}$

Step 3   The first term a_1 is 216 and $n \geq 2$.

A recursive formula for the sequence is $a_1 = 216$, $a_n = \frac{1}{6} a_{n-1}$, $n \geq 2$.

Exercises

Write a recursive formula for each sequence.

1. 22, 16, 10, 4, ...

2. −8, −3, 2, 7, ...

3. 5, 15, 45, 135, ...

4. 243, 81, 27, 9, ...

5. −3, 14, 31, 48, ...

6. 8, −20, 50, −125, ...

8-1 Study Guide and Intervention

Adding and Subtracting Polynomials

Polynomials in Standard Form A **polynomial** is a monomial or a sum of monomials. A **binomial** is the sum of two monomials, and a **trinomial** is the sum of three monomials. Polynomials with more than three terms have no special name. The **degree** of a monomial is the sum of the exponents of all its variables. The **degree of the polynomial** is the same as the degree of the monomial term with the highest degree.

The terms of a polynomial are usually arranged so that the terms are in order from greatest degree to least degree. This is called the **standard form of a polynomial.**

Example **Determine whether each expression is a polynomial. If so, identify the polynomial as a** *monomial,* *binomial,* **or** *trinomial.* **Then find the degree of the polynomial.**

Expression	Polynomial?	Monomial, Binomial, or Trinomial?	Degree of the Polynomial
$3x - 7xyz$	Yes. $3x - 7xyz = 3x + (-7xyz)$, which is the sum of two monomials	binomial	3
-25	Yes. -25 is a real number.	monomial	0
$7n^3 + 3n^{-4}$	No. $3n^{-4} = \dfrac{3}{n^4}$, which is not a monomial	none of these	—
$9x^3 + 4x + x + 4 + 2x$	Yes. The expression simplifies to $9x^3 + 7x + 4$, which is the sum of three monomials	trinomial	3

Exercises

Determine whether each expression is a polynomial. If it is a polynomial, find the degree and determine whether it is a *monomial,* *binomial,* **or** *trinomial.*

1. 36

2. $\dfrac{3}{q^2} + 5$

3. $7x - x + 5$

4. $8g^2h - 7gh + 2$

5. $\dfrac{1}{4y^2} + 5y - 8$

6. $6x + x^2$

Write each polynomial in standard form. Identify the leading coefficient.

7. $x^3 + x^5 - x^2$

8. $x^4 + 4x^3 - 7x^5 + 1$

9. $-3x^6 - x^5 + 2x^8$

10. $2x^7 - x^8$

11. $3x + 5x^4 - 2 - x^2$

12. $-2x^4 + x - 4x^5 + 3$

8-1   Study Guide and Intervention (continued)

Adding and Subtracting Polynomials

Add and Subtract Polynomials To add polynomials, you can group like terms horizontally or write them in column form, aligning like terms vertically. **Like terms** are monomial terms that are either identical or differ only in their coefficients, such as $3p$ and $-5p$ or $2x^2y$ and $8x^2y$.

You can subtract a polynomial by adding its additive inverse. To find the additive inverse of a polynomial, replace each term with by adding its additive inverse. To find the additive inverse of a polynomial, replace each term with its additive inverse or opposite.

> **Example** Find $(3x^2 + 2x - 6) - (2x + x^2 + 3)$.

Horizontal Method

Use additive inverses to rewrite as addition. Then group like terms.

$(3x^2 + 2x - 6) - (2x + x^2 + 3)$
$= (3x^2 + 2x - 6) + [(-2x) + (-x^2) + (-3)]$
$= [3x^2 + (-x^2)] + [2x + (-2x)] + [-6 + (-3)]$
$= 2x^2 + (-9)$
$= 2x^2 - 9$

The difference is $2x^2 - 9$.

Vertical Method

Align like terms in columns and subtract by adding the additive inverse.

$$\begin{array}{r} 3x^2 + 2x - 6 \\ \hline (-)\ \ x^2 + 2x + 3 \end{array}$$

$$\begin{array}{r} 3x^2 + 2x - 6 \\ \hline (+) -x^2 - 2x - 3 \\ \hline 2x^2 \qquad\ \ - 9 \end{array}$$

The difference is $2x^2 - 9$.

Exercises

Find each sum or difference.

1. $(4a - 5) + (3a + 6)$

2. $(6x + 9) + (4x^2 - 7)$

3. $(6xy + 2y + 6x) + (4xy - x)$

4. $(x^2 + y^2) + (-x^2 + y^2)$

5. $(3p^2 - 2p + 3) + (p^2 - 7p + 7)$

6. $(2x^2 + 5xy + 4y^2) + (-xy - 6x^2 + 2y^2)$

7. $(8p - 5r) - (-6p^2 + 6r - 3)$

8. $(8x^2 - 4x - 3) - (-2x - x^2 + 5)$

9. $(3x^2 - 2x) - (3x^2 + 5x - 1)$

10. $(4x^2 + 6xy + 2y^2) - (-x^2 + 2xy - 5y^2)$

11. $(2h - 6j - 2k) - (-7h - 5j - 4k)$

12. $(9xy^2 + 5xy) - (-2xy - 8xy^2)$

8-2 **Study Guide and Intervention**

Multiplying a Polynomial by a Monomial

Polynomial Multiplied by Monomial The Distributive Property can be used to multiply a polynomial by a monomial. You can multiply horizontally or vertically. Sometimes multiplying results in like terms. The products can be simplified by combining like terms.

Example 1 Find $-3x^2(4x^2 + 6x - 8)$.

Horizontal Method

$-3x^2(4x^2 + 6x - 8)$
$= -3x^2(4x^2) + (-3x^2)(6x) - (-3x^2)(8)$
$= -12x^4 + (-18x^3) - (-24x^2)$
$= -12x^4 - 18x^3 + 24x^2$

Vertical Method

$$
\begin{array}{r}
4x^2 + 6x - 8 \\
(\times) \qquad\qquad -3x^2 \\
\hline
-12x^4 - 18x^3 + 24x^2
\end{array}
$$

The product is $-12x^4 - 18x^3 + 24x^2$.

Example 2 Simplify $-2(4x^2 + 5x) - x(x^2 + 6x)$.

$-2(4x^2 + 5x) - x(x^2 + 6x)$
$= -2(4x^2) + (-2)(5x) + (-x)(x^2) + (-x)(6x)$
$= -8x^2 + (-10x) + (-x^3) + (-6x^2)$
$= (-x^3) + [-8x^2 + (-6x^2)] + (-10x)$
$= -x^3 - 14x^2 - 10x$

Exercises

Find each product.

1. $x(5x + x^2)$

2. $x(4x^2 + 3x + 2)$

3. $-2xy(2y + 4x^2)$

4. $-2g(g^2 - 2g + 2)$

5. $3x(x^4 + x^3 + x^2)$

6. $-4x(2x^3 - 2x + 3)$

7. $-4ax(10 + 3x)$

8. $3y(-4x - 6x^3 - 2y)$

9. $2x^2y^2(3xy + 2y + 5x)$

Simplify each expression.

10. $x(3x - 4) - 5x$

11. $-x(2x^2 - 4x) - 6x^2$

12. $6a(2a - b) + 2a(-4a + 5b)$

13. $4r(2r^2 - 3r + 5) + 6r(4r^2 + 2r + 8)$

14. $4n(3n^2 + n - 4) - n(3 - n)$

15. $2b(b^2 + 4b + 8) - 3b(3b^2 + 9b - 18)$

16. $-2z(4z^2 - 3z + 1) - z(3z^2 + 2z - 1)$

17. $2(4x^2 - 2x) - 3(-6x^2 + 4) + 2x(x - 1)$

8-2 Study Guide and Intervention (continued)

Multiplying a Polynomial by a Monomial

Solve Equations with Polynomial Expressions Many equations contain polynomials that must be added, subtracted, or multiplied before the equation can be solved.

Example Solve $4(n - 2) + 5n = 6(3 - n) + 19$.

$4(n - 2) + 5n = 6(3 - n) + 19$	Original equation
$4n - 8 + 5n = 18 - 6n + 19$	Distributive Property
$9n - 8 = 37 - 6n$	Combine like terms.
$15n - 8 = 37$	Add $6n$ to both sides.
$15n = 45$	Add 8 to both sides.
$n = 3$	Divide each side by 15.

The solution is 3.

Exercises

Solve each equation.

1. $2(a - 3) = 3(-2a + 6)$

2. $3(x + 5) - 6 = 18$

3. $3x(x - 5) - 3x^2 = -30$

4. $6(x^2 + 2x) = 2(3x^2 + 12)$

5. $4(2p + 1) - 12p = 2(8p + 12)$

6. $2(6x + 4) + 2 = 4(x - 4)$

7. $-2(4y - 3) - 8y + 6 = 4(y - 2)$

8. $x(x + 2) - x(x - 6) = 10x - 12$

9. $3(x^2 - 2x) = 3x^2 + 5x - 11$

10. $2(4x + 3) + 2 = -4(x + 1)$

11. $3(2h - 6) - (2h + 1) = 9$

12. $3(y + 5) - (4y - 8) = -2y + 10$

13. $3(2a - 6) - (-3a - 1) = 4a - 2$

14. $5(2x^2 - 1) - (10x^2 - 6) = -(x + 2)$

15. $3(x + 2) + 2(x + 1) = -5(x - 3)$

16. $4(3p^2 + 2p) - 12p^2 = 2(8p + 6)$

8-3 Study Guide and Intervention

Multiplying Polynomials

Multiply Binomials To multiply two binomials, you can apply the Distributive Property twice. A useful way to keep track of terms in the product is to use the FOIL method as illustrated in Example 2.

Example 1 Find $(x + 3)(x - 4)$.

Horizontal Method

$(x + 3)(x - 4)$
$= x(x - 4) + 3(x - 4)$
$= (x)(x) + x(-4) + 3(x) + 3(-4)$
$= x^2 - 4x + 3x - 12$
$= x^2 - x - 12$

Vertical Method

$\quad\quad x + 3$
$(\times)\quad x - 4$
$\overline{\quad -4x - 12}$
$\underline{x^2 + 3x}$
$x^2 - \;x - 12$

The product is $x^2 - x - 12$.

Example 2 Find $(x - 2)(x + 5)$ using the FOIL method.

$(x - 2)(x + 5)$

$\quad$ First $\quad$ Outer $\quad$ Inner $\quad$ Last
$= (x)(x) + (x)(5) + (-2)(x) + (-2)(5)$
$= x^2 + 5x + (-2x) - 10$
$= x^2 + 3x - 10$

The product is $x^2 + 3x - 10$.

Exercises

Find each product.

1. $(x + 2)(x + 3)$

2. $(x - 4)(x + 1)$

3. $(x - 6)(x - 2)$

4. $(p - 4)(p + 2)$

5. $(y + 5)(y + 2)$

6. $(2x - 1)(x + 5)$

7. $(3n - 4)(3n - 4)$

8. $(8m - 2)(8m + 2)$

9. $(k + 4)(5k - 1)$

10. $(3x + 1)(4x + 3)$

11. $(x - 8)(-3x + 1)$

12. $(5t + 4)(2t - 6)$

13. $(5m - 3n)(4m - 2n)$

14. $(a - 3b)(2a - 5b)$

15. $(8x - 5)(8x + 5)$

16. $(2n - 4)(2n + 5)$

17. $(4m - 3)(5m - 5)$

18. $(7g - 4)(7g + 4)$

8-3 Study Guide and Intervention (continued)

Multiplying Polynomials

Multiply Polynomials The Distributive Property can be used to multiply any two polynomials.

Example Find $(3x + 2)(2x^2 - 4x + 5)$.

$(3x + 2)(2x^2 - 4x + 5)$

$= 3x(2x^2 - 4x + 5) + 2(2x^2 - 4x + 5)$ Distributive Property

$= 6x^3 - 12x^2 + 15x + 4x^2 - 8x + 10$ Distributive Property

$= 6x^3 - 8x^2 + 7x + 10$ Combine like terms.

The product is $6x^3 - 8x^2 + 7x + 10$.

Exercises

Find each product.

1. $(x + 2)(x^2 - 2x + 1)$

2. $(x + 3)(2x^2 + x - 3)$

3. $(2x - 1)(x^2 - x + 2)$

4. $(p - 3)(p^2 - 4p + 2)$

5. $(3k + 2)(k^2 + k - 4)$

6. $(2t + 1)(10t^2 - 2t - 4)$

7. $(3n - 4)(n^2 + 5n - 4)$

8. $(8x - 2)(3x^2 + 2x - 1)$

9. $(2a + 4)(2a^2 - 8a + 3)$

10. $(3x - 4)(2x^2 + 3x + 3)$

11. $(n^2 + 2n - 1)(n^2 + n + 2)$

12. $(t^2 + 4t - 1)(2t^2 - t - 3)$

13. $(y^2 - 5y + 3)(2y^2 + 7y - 4)$

14. $(3b^2 - 2b + 1)(2b^2 - 3b - 4)$

8-4 Study Guide and Intervention

Special Products

Squares of Sums and Differences Some pairs of binomials have products that follow specific patterns. One such pattern is called the *square of a sum*. Another is called the *square of a difference*.

Square of a Sum	$(a + b)^2 = (a + b)(a + b) = a^2 + 2ab + b^2$
Square of a Difference	$(a - b)^2 = (a - b)(a - b) = a^2 - 2ab + b^2$

Example 1 Find $(3a + 4)(3a + 4)$.

Use the square of a sum pattern, with $a = 3a$ and $b = 4$.

$(3a + 4)(3a + 4) = (3a)^2 + 2(3a)(4) + (4)^2$
$\qquad\qquad\qquad = 9a^2 + 24a + 16$

The product is $9a^2 + 24a + 16$.

Example 2 Find $(2z - 9)(2z - 9)$.

Use the square of a difference pattern with $a = 2z$ and $b = 9$.

$(2z - 9)(2z - 9) = (2z)^2 - 2(2z)(9) + (9)(9)$
$\qquad\qquad\qquad = 4z^2 - 36z + 81$

The product is $4z^2 - 36z + 81$.

Exercises

Find each product.

1. $(x - 6)^2$

2. $(3p + 4)^2$

3. $(4x - 5)^2$

4. $(2x - 1)^2$

5. $(2h + 3)^2$

6. $(m + 5)^2$

7. $(a + 3)^2$

8. $(3 - p)^2$

9. $(x - 5y)^2$

10. $(8y + 4)^2$

11. $(8 + x)^2$

12. $(3a - 2b)^2$

13. $(2x - 8)^2$

14. $(x^2 + 1)^2$

15. $(m^2 - 2)^2$

16. $(x^3 - 1)^2$

17. $(2h^2 - k^2)^2$

18. $\left(\dfrac{1}{4}x + 3\right)^2$

19. $(x - 4y^2)^2$

20. $(2p + 4r)^2$

21. $\left(\dfrac{2}{3}x - 2\right)^2$

8-4　Study Guide and Intervention (continued)

Special Products

Product of a Sum and a Difference There is also a pattern for the product of a sum and a difference of the same two terms, $(a + b)(a - b)$. The product is called the **difference of squares**.

Product of a Sum and a Difference	$(a + b)(a - b) = a^2 - b^2$

Example　Find $(5x + 3y)(5x - 3y)$.

$(a + b)(a - b) = a^2 - b^2$　　　Product of a Sum and a Difference

$(5x + 3y)(5x - 3y) = (5x)^2 - (3y)^2$　　　$a = 5x$ and $b = 3y$

$= 25x^2 - 9y^2$　　　Simplify.

The product is $25x^2 - 9y^2$.

Exercises

Find each product.

1. $(x - 4)(x + 4)$　　　**2.** $(p + 2)(p - 2)$　　　**3.** $(4x - 5)(4x + 5)$

4. $(2x - 1)(2x + 1)$　　　**5.** $(h + 7)(h - 7)$　　　**6.** $(m - 5)(m + 5)$

7. $(2d - 3)(2d + 3)$　　　**8.** $(3 - 5q)(3 + 5q)$　　　**9.** $(x - y)(x + y)$

10. $(y - 4x)(y + 4x)$　　　**11.** $(8 + 4x)(8 - 4x)$　　　**12.** $(3a - 2b)(3a + 2b)$

13. $(3y - 8)(3y + 8)$　　　**14.** $(x^2 - 1)(x^2 + 1)$　　　**15.** $(m^2 - 5)(m^2 + 5)$

16. $(x^3 - 2)(x^3 + 2)$　　　**17.** $(h^2 - k^2)(h^2 + k^2)$　　　**18.** $\left(\frac{1}{4}x + 2\right)\left(\frac{1}{4}x - 2\right)$

19. $(3x - 2y^2)(3x + 2y^2)$　　　**20.** $(2p - 5r)(2p + 5r)$　　　**21.** $\left(\frac{4}{3}x - 2y\right)\left(\frac{4}{3}x + 2y\right)$

8-5 Study Guide and Intervention

Using the Distributive Property

Use the Distributive Property to Factor The Distributive Property has been used to multiply a polynomial by a monomial. It can also be used to express a polynomial in factored form. Compare the two columns in the table below.

Multiplying	Factoring
$3(a + b) = 3a + 3b$	$3a + 3b = 3(a + b)$
$x(y - z) = xy - xz$	$xy - xz = x(y - z)$
$6y(2x + 1) = 6y(2x) + 6y(1)$ $\quad = 12xy + 6y$	$12xy + 6y = 6y(2x) + 6y(1)$ $\quad = 6y(2x + 1)$

Example 1 Use the Distributive Property to factor $12mp + 80m^2$.

Find the GCF of $12mp$ and $80m^2$.

$12mp = 2 \cdot 2 \cdot 3 \cdot m \cdot p$
$80m^2 = 2 \cdot 2 \cdot 2 \cdot 2 \cdot 5 \cdot m \cdot m$
$GCF = 2 \cdot 2 \cdot m$ or $4m$

Write each term as the product of the GCF and its remaining factors.

$12mp + 80m^2 = 4m(3 \cdot p) + 4m(2 \cdot 2 \cdot 5 \cdot m)$
$\qquad\qquad = 4m(3p) + 4m(20m)$
$\qquad\qquad = 4m(3p + 20m)$

Thus $12mp + 80m^2 = 4m(3p + 20m)$.

Example 2 Factor $6ax + 3ay + 2bx + by$ by grouping.

$6ax + 3ay + 2bx + by$
$\quad = (6ax + 3ay) + (2bx + by)$
$\quad = 3a(2x + y) + b(2x + y)$
$\quad = (3a + b)(2x + y)$

Check using the FOIL method.
$(3a + b)(2x + y)$
$\quad = 3a(2x) + (3a)(y) + (b)(2x) + (b)(y)$
$\quad = 6ax + 3ay + 2bx + by$ ✓

Exercises

Factor each polynomial.

1. $24x + 48y$

2. $30mp^2 + m^2p - 6p$

3. $q^4 - 18q^3 + 22q$

4. $9x^2 - 3x$

5. $4m + 6p - 8mp$

6. $45r^3 - 15r^2$

7. $14t^3 - 42t^5 - 49t^4$

8. $55p^2 - 11p^4 + 44p^5$

9. $14y^3 - 28y^2 + y$

10. $4x + 12x^2 + 16x^3$

11. $4a^2b + 28ab^2 + 7ab$

12. $6y + 12x - 8z$

13. $x^2 + 2x + x + 2$

14. $6y^2 - 4y + 3y - 2$

15. $4m^2 + 4mp + 3mp + 3p^2$

16. $12ax + 3xz + 4ay + yz$

17. $12a^2 + 3a - 8a - 2$

18. $xa + ya + x + y$

8-5 Study Guide and Intervention (continued)

Using the Distributive Property

Solve Equations by Factoring The following property, along with factoring, can be used to solve certain equations.

Zero Product Property	For any real numbers a and b, if $ab = 0$, then either $a = 0$, $b = 0$, or both a and b equal 0.

Example **Solve $9x^2 + x = 0$. Then check the solutions.**

Write the equation so that it is of the form $ab = 0$.

$9x^2 + x = 0$	Original equation
$x(9x + 1) = 0$	Factor the GCF of $9x^2 + x$, which is x.
$x = 0$ or $9x + 1 = 0$	Zero Product Property
$x = 0 \qquad x = -\dfrac{1}{9}$	Solve each equation.

The solution set is $\left\{0, -\dfrac{1}{9}\right\}$.

Check Substitute 0 and $-\dfrac{1}{9}$ for x in the original equation.

$$9x^2 + x = 0 \qquad\qquad 9x^2 + x = 0$$
$$9(0)^2 + 0 \stackrel{?}{=} 0 \qquad 9\left(-\dfrac{1}{9}\right)^2 + \left(-\dfrac{1}{9}\right) \stackrel{?}{=} 0$$
$$0 = 0 \checkmark \qquad\qquad \dfrac{1}{9} + \left(-\dfrac{1}{9}\right) \stackrel{?}{=} 0$$
$$0 = 0 \checkmark$$

Exercises

Solve each equation. Check your solutions.

1. $x(x + 3) = 0$

2. $3m(m - 4) = 0$

3. $(r - 3)(r + 2) = 0$

4. $3x(2x - 1) = 0$

5. $(4m + 8)(m - 3) = 0$

6. $5t^2 = 25t$

7. $(4c + 2)(2c - 7) = 0$

8. $5p - 15p^2 = 0$

9. $4y^2 = 28y$

10. $12x^2 = -6x$

11. $(4a + 3)(8a + 7) = 0$

12. $8y = 12y^2$

13. $x^2 = -2x$

14. $(6y - 4)(y + 3) = 0$

15. $4m^2 = 4m$

16. $12x = 3x^2$

17. $12a^2 = -3a$

18. $(12a + 4)(3a - 1) = 0$

8-6 Study Guide and Intervention

Solving $x^2 + bx + c = 0$

Factor $x^2 + bx + c$ To factor a trinomial of the form $x^2 + bx + c$, find two integers, m and p, whose sum is equal to b and whose product is equal to c.

Factoring $x^2 + bx + c$	$x^2 + bx + c = (x + m)(x + p)$, where $m + p = b$ and $mp = c$.

Example 1 Factor each polynomial.

a. $x^2 + 7x + 10$

In this trinomial, $b = 7$ and $c = 10$.

Factors of 10	Sum of Factors
1, 10	11
2, 5	7

Since $2 + 5 = 7$ and $2 \cdot 5 = 10$, let $m = 2$ and $p = 5$.

$x^2 + 7x + 10 = (x + 5)(x + 2)$

b. $x^2 - 8x + 7$

In this trinomial, $b = -8$ and $c = 7$.

Notice that $m + p$ is negative and mp is positive, so m and p are both negative.
Since $-7 + (-1) = -8$ and $(-7)(-1) = 7$, $m = -7$ and $p = -1$.

$x^2 - 8x + 7 = (x - 7)(x - 1)$

Example 2 Factor $x^2 + 6x - 16$.

In this trinomial, $b = 6$ and $c = -16$. This means $m + p$ is positive and mp is negative. Make a list of the factors of -16, where one factor of each pair is positive.

Factors of −16	Sum of Factors
1, −16	−15
−1, 16	15
2, −8	−6
−2, 8	6

Therefore, $m = -2$ and $p = 8$.

$x^2 + 6x - 16 = (x - 2)(x + 8)$

Exercises

Factor each polynomial.

1. $x^2 + 4x + 3$

2. $m^2 + 12m + 32$

3. $r^2 - 3r + 2$

4. $x^2 - x - 6$

5. $x^2 - 4x - 21$

6. $x^2 - 22x + 121$

7. $t^2 - 4t - 12$

8. $p^2 - 16p + 64$

9. $9 - 10x + x^2$

10. $x^2 + 6x + 5$

11. $a^2 + 8a - 9$

12. $y^2 - 7y - 8$

13. $x^2 - 2x - 3$

14. $y^2 + 14y + 13$

15. $m^2 + 9m + 20$

16. $x^2 + 12x + 20$

17. $a^2 - 14a + 24$

18. $18 + 11y + y^2$

19. $x^2 + 2xy + y^2$

20. $a^2 - 4ab + 4b^2$

21. $x^2 + 6xy - 7y^2$

8-6 Study Guide and Intervention (continued)

Solving $x^2 + bx + c = 0$

Solve Equations by Factoring Factoring and the Zero Product Property can be used to solve many equations of the form $x^2 + bx + c = 0$.

Example 1 Solve $x^2 + 6x = 7$. Check your solutions.

$x^2 + 6x = 7$	Original equation
$x^2 + 6x - 7 = 0$	Rewrite equation so that one side equals 0.
$(x - 1)(x + 7) = 0$	Factor.
$x - 1 = 0$ or $x + 7 = 0$	Zero Product Property
$x = 1$ $x = -7$	Solve each equation.

The solution set is $\{1, -7\}$. Since $1^2 + 6(1) = 7$ and $(-7)^2 + 6(-7) = 7$, the solutions check.

Example 2 ROCKET LAUNCH The formula $h = vt - 16t^2$ gives the height h of a rocket after t seconds when the initial velocity v is given in feet per second. If a rocket is fired with initial velocity 2288 feet per second, how many seconds will it take for the rocket to reach a height of 6720 feet?

$h = vt - 16t^2$	Formula
$6720 = 2288t - 16t^2$	Substitute.
$0 = -16t^2 + 2288t - 6720$	Rewrite equation so that one side equals 0.
$0 = -16(t - 143t + 420)$	Factor out GCF.
$0 = -16(t - 3)(t - 140)$	Factor
$t - 3 = 0$ or $t - 140 = 0$	Zero Product Property
$t = 3$ $t = 140$	Solve each equation.

The rocket reaches 6720 feet in 3 seconds and again in 140 seconds, or 2 minutes 20 seconds after launch.

Exercises

Solve each equation. Check the solutions.

1. $x^2 - 4x + 3 = 0$　　　　**2.** $y^2 - 5y + 4 = 0$　　　　**3.** $m^2 + 10m + 9 = 0$

4. $x^2 = x + 2$　　　　　　　**5.** $x^2 - 4x = 5$　　　　　　**6.** $x^2 - 12x + 36 = 0$

7. $t^2 - 8 = -7t$　　　　　　**8.** $p^2 = 9p - 14$　　　　　**9.** $-9 - 8x + x^2 = 0$

10. $x^2 + 6 = 5x$　　　　　　**11.** $a^2 = 11a - 18$　　　　**12.** $y^2 - 8y + 15 = 0$

13. $x^2 = 24 - 10x$　　　　　**14.** $a^2 - 18a = -72$　　　**15.** $b^2 = 10b - 16$

Use the formula $h = vt - 16t^2$ to solve each problem.

16. FOOTBALL A punter can kick a football with an initial velocity of 48 feet per second. How many seconds will it take for the ball to first reach a height of 32 feet?

17. ROCKET LAUNCH If a rocket is launched with an initial velocity of 1600 feet per second, when will the rocket be 14,400 feet high?

8-7   Study Guide and Intervention
Solving $ax^2 + bx + c = 0$

Factor $ax^2 + bx + c$ To factor a trinomial of the form $ax^2 + bx + c$, find two integers, m and p whose product is equal to ac and whose sum is equal to b. If there are no integers that satisfy these requirements, the polynomial is called a **prime polynomial**.

Example 1   **Factor $2x^2 + 15x + 18$.**

In this example, $a = 2$, $b = 15$, and $c = 18$. You need to find two numbers that have a sum of 15 and a product of $2 \cdot 18$ or 36. Make a list of the factors of 36 and look for the pair of factors with a sum of 15.

Factors of 36	Sum of Factors
1, 36	37
2, 18	20
3, 12	15

Use the pattern $ax^2 + mx + px + c$, with $a = 2$, $m = 3$, $p = 12$, and $c = 18$.

$$2x^2 + 15x + 18 = 2x^2 + 3x + 12x + 18$$
$$= (2x^2 + 3x) + (12x + 18)$$
$$= x(2x + 3) + 6(2x + 3)$$
$$= (x + 6)(2x + 3)$$

Therefore, $2x^2 + 15x + 18 = (x + 6)(2x + 3)$.

Example 2   **Factor $3x^2 - 3x - 18$.**

Note that the GCF of the terms $3x^2$, $3x$, and 18 is 3. First factor out this GCF.
$$3x^2 - 3x - 18 = 3(x^2 - x - 6).$$
Now factor $x^2 - x - 6$. Since $a = 1$, find the two factors of -6 with a sum of -1.

Factors of -6	Sum of Factors
1, -6	-5
-1, 6	5
-2, 3	1
2, -3	-1

Now use the pattern $(x + m)(x + p)$ with $m = 2$ and $p = -3$.
$$x^2 - x - 6 = (x + 2)(x - 3)$$
The complete factorization is
$$3x^2 - 3x - 18 = 3(x + 2)(x - 3).$$

Exercises

Factor each polynomial, if possible. If the polynomial cannot be factored using integers, write *prime*.

1. $2x^2 - 3x - 2$

2. $3m^2 - 8m - 3$

3. $16r^2 - 8r + 1$

4. $6x^2 + 5x - 6$

5. $3x^2 + 2x - 8$

6. $18x^2 - 27x - 5$

7. $2a^2 + 5a + 3$

8. $18y^2 + 9y - 5$

9. $-4t^2 + 19t - 21$

10. $8x^2 - 4x - 24$

11. $28p^2 + 60p - 25$

12. $48x^2 + 22x - 15$

13. $3y^2 - 6y - 24$

14. $4x^2 + 26x - 48$

15. $8m^2 - 44m + 48$

16. $6x^2 - 7x + 18$

17. $2a^2 - 14a + 18$

18. $18 + 11y + 2y^2$

8-7 **Study Guide and Intervention** (continued)

Solving $ax^2 + bx + c = 0$

Solve Equations by Factoring Factoring and the Zero Product Property can be used to solve some equations of the form $ax^2 + bx + c = 0$.

Example **Solve $12x^2 + 3x = 2 - 2x$. Check your solutions.**

$12x^2 + 3x = 2 - 2x$	Original equation
$12x^2 + 5x - 2 = 0$	Rewrite equation so that one side equals 0.
$(3x + 2)(4x - 1) = 0$	Factor the left side.
$3x + 2 = 0$ or $4x - 1 = 0$	Zero Product Property
$x = -\dfrac{2}{3}$ $\qquad x = \dfrac{1}{4}$	Solve each equation.

The solution set is $\left\{-\dfrac{2}{3}, \dfrac{1}{4}\right\}$.

Since $12\left(-\dfrac{2}{3}\right)^2 + 3\left(-\dfrac{2}{3}\right) = 2 - 2\left(-\dfrac{2}{3}\right)$ and $12\left(\dfrac{1}{4}\right)^2 + 3\left(\dfrac{1}{4}\right) = 2 - 2\left(\dfrac{1}{4}\right)$, the solutions check.

Exercises

Solve each equation. Check the solutions.

1. $8x^2 + 2x - 3 = 0$

2. $3n^2 - 2n - 5 = 0$

3. $2d^2 - 13d - 7 = 0$

4. $4x^2 = x + 3$

5. $3x^2 - 13x = 10$

6. $6x^2 - 11x - 10 = 0$

7. $2k^2 - 40 = -11k$

8. $2p^2 = -21p - 40$

9. $-7 - 18x + 9x^2 = 0$

10. $12x^2 - 15 = -8x$

11. $7a^2 = -65a - 18$

12. $16y^2 - 2y - 3 = 0$

13. $8x^2 + 5x = 3 + 7x$

14. $4a^2 - 18a + 5 = 15$

15. $3b^2 - 18b = 10b - 49$

16. The difference of the squares of two consecutive odd integers is 24. Find the integers.

17. GEOMETRY The length of a Charlotte, North Carolina, conservatory garden is 20 yards greater than its width. The area is 300 square yards. What are the dimensions?

18. GEOMETRY A rectangle with an area of 24 square inches is formed by cutting strips of equal width from a rectangular piece of paper. Find the dimensions of the new rectangle if the original rectangle measures 8 inches by 6 inches.

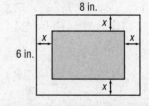

8-8 Study Guide and Intervention

Differences of Squares

Factor Differences of Squares The binomial expression $a^2 - b^2$ is called the **difference of two squares**. The following pattern shows how to factor the difference of squares.

Difference of Squares	$a^2 - b^2 = (a - b)(a + b) = (a + b)(a - b)$.

Example 1 Factor each polynomial.

a. $n^2 - 64$

$n^2 - 64$

$= n^2 - 8^2$ Write in the form $a^2 - b^2$.

$= (n + 8)(n - 8)$ Factor.

b. $4m^2 - 81n^2$

$4m^2 - 81n^2$

$= (2m)^2 - (9n)^2$ Write in the form $a^2 - b^2$.

$= (2m - 9n)(2m + 9n)$ Factor.

Example 2 Factor each polynomial.

a. $50a^2 - 72$

$50a^2 - 72$

$= 2(25a^2 - 36)$ Find the GCF.

$= 2[(5a)^2 - 6^2)]$ $25a^2 = 5a \cdot 5a$ and $36 = 6 \cdot 6$

$= 2(5a + 6)(5a - 6)$ Factor the difference of squares.

b. $4x^4 + 8x^3 - 4x^2 - 8x$

$4x^4 + 8x^3 - 4x^2 - 8x$ Original polynomial

$= 4x(x^3 + 2x^2 - x - 2)$ Find the GCF.

$= 4x[(x^3 + 2x^2) - (x + 2)]$ Group terms.

$= 4x[x^2(x + 2) - 1(x + 2)]$ Find the GCF.

$= 4x[(x^2 - 1)(x + 2)]$ Factor by grouping.

$= 4x[(x - 1)(x + 1)(x + 2)]$ Factor the difference of squares.

Exercises

Factor each polynomial.

1. $x^2 - 81$

2. $m^2 - 100$

3. $16n^2 - 25$

4. $36x^2 - 100y^2$

5. $49x^2 - 36$

6. $16a^2 - 9b^2$

7. $225b^2 - a^2$

8. $72p^2 - 50$

9. $-2 + 2x^2$

10. $-81 + a^4$

11. $6 - 54a^2$

12. $8y^2 - 200$

13. $4x^3 - 100x$

14. $2y^4 - 32y^2$

15. $8m^3 - 128m$

16. $4x^2 - 25$

17. $2a^3 - 98ab^2$

18. $18y^2 - 72y^4$

19. $169x^3 - x$

20. $3a^4 - 3a^2$

21. $3x^4 + 6x^3 - 3x^2 - 6x$

8-8 Study Guide and Intervention *(continued)*

Differences of Squares

Solve Equations by Factoring Factoring and the Zero Product Property can be used to solve equations that can be written as the product of any number of factors set equal to 0.

Example Solve each equation. Check your solutions.

a. $x^2 - \frac{1}{25} = 0$

$$x^2 - \frac{1}{25} = 0 \qquad \text{Original equation}$$

$$x^2 - \left(\frac{1}{5}\right)^2 = 0 \qquad x^2 = x \cdot x \text{ and } \frac{1}{25} = \left(\frac{1}{5}\right)\left(\frac{1}{5}\right)$$

$$\left(x + \frac{1}{5}\right)\left(x - \frac{1}{5}\right) = 0 \qquad \text{Factor the difference of squares.}$$

$$x + \frac{1}{5} = 0 \quad \text{or} \quad x - \frac{1}{5} = 0 \qquad \text{Zero Product Property}$$

$$x = -\frac{1}{5} \qquad\qquad x = \frac{1}{5} \qquad \text{Solve each equation.}$$

The solution set is $\left\{-\frac{1}{5}, \frac{1}{5}\right\}$. Since $\left(-\frac{1}{5}\right)^2 - \frac{1}{25} = 0$ and $\left(\frac{1}{5}\right)^2 - \frac{1}{25} = 0$, the solutions check.

b. $4x^3 = 9x$

$$4x^3 = 9x \qquad \text{Original equation}$$

$$4x^3 - 9x = 0 \qquad \text{Subtract } 9x \text{ from each side.}$$

$$x(4x^2 - 9) = 0 \qquad \text{Factor out the GCF of } x.$$

$$x[(2x)^2 - 3^2] = 0 \qquad 4x^2 = 2x \cdot 2x \text{ and } 9 = 3 \cdot 3$$

$$x[(2x)^2 - 3^2] = x[(2x - 3)(2x + 3)] \qquad \text{Factor the difference of squares.}$$

$$x = 0 \quad \text{or} \quad (2x - 3) = 0 \quad \text{or} \quad (2x + 3) = 0 \qquad \text{Zero Product Property}$$

$$x = 0 \qquad\qquad x = \frac{3}{2} \qquad\qquad x = -\frac{3}{2} \qquad \text{Solve each equation.}$$

The solution set is $\left\{0, \frac{3}{2}, -\frac{3}{2}\right\}$.

Since $4(0)^3 = 9(0)$, $4\left(\frac{3}{2}\right)^3 = 9\left(\frac{3}{2}\right)$, and $4\left(-\frac{3}{2}\right)^3 = 9\left(-\frac{3}{2}\right)$, the solutions check.

Exercises

Solve each equation by factoring. Check the solutions.

1. $81x^2 = 49$

2. $36n^2 = 1$

3. $25d^2 - 100 = 0$

4. $\frac{1}{4}x^2 = 25$

5. $36 = \frac{1}{25}x^2$

6. $\frac{49}{100} - x^2 = 0$

7. $9x^3 = 25x$

8. $7a^3 = 175a$

9. $2m^3 = 32m$

10. $16y^3 = 25y$

11. $\frac{1}{64}x^2 = 49$

12. $4a^3 - 64a = 0$

13. $3b^3 - 27b = 0$

14. $\frac{9}{25}m^2 = 121$

15. $48n^3 = 147n$

8-9 Study Guide and Intervention

Perfect Squares

Factor Perfect Square Trinomials

Perfect Square Trinomial	a trinomial of the form $a^2 + 2ab + b^2$ or $a^2 - 2ab + b^2$

The patterns shown below can be used to factor perfect square trinomials.

Squaring a Binomial	Factoring a Perfect Square Trinomial
$(a + 4)^2 = a^2 + 2(a)(4) + 4^2$ $= a^2 + 8a + 16$	$a^2 + 8a + 16 = a^2 + 2(a)(4) + 4^2$ $= (a + 4)^2$
$(2x - 3)^2 = (2x)^2 - 2(2x)(3) + 3^2$ $= 4x^2 - 12x + 9$	$4x^2 - 12x + 9 = (2x)^2 - 2(2x)(3) + 3^2$ $= (2x - 3)^2$

Example 1 Determine whether $16n^2 - 24n + 9$ is a perfect square trinomial. If so, factor it.

Since $16n^2 = (4n)(4n)$, the first term is a perfect square.

Since $9 = 3 \cdot 3$, the last term is a perfect square.

The middle term is equal to $2(4n)(3)$.

Therefore, $16n^2 - 24n + 9$ is a perfect square trinomial.

$16n^2 - 24n + 9 = (4n)^2 - 2(4n)(3) + 3^2$
$= (4n - 3)^2$

Example 2 Factor $16x^2 - 32x + 15$.

Since 15 is not a perfect square, use a different factoring pattern.

$16x^2 - 32x + 15$ Original trinomial
$= 16x^2 + mx + px + 15$ Write the pattern.
$= 16x^2 - 12x - 20x + 15$ $m = -12$ and $p = -20$
$= (16x^2 - 12x) - (20x - 15)$ Group terms.
$= 4x(4x - 3) - 5(4x - 3)$ Find the GCF.
$= (4x - 5)(4x - 3)$ Factor by grouping.

Therefore $16x^2 - 32x + 15 = (4x - 5)(4x - 3)$.

Exercises

Determine whether each trinomial is a perfect square trinomial. Write *yes* or *no*. If so, factor it.

1. $x^2 - 16x + 64$ **2.** $m^2 + 10m + 25$ **3.** $p^2 + 8p + 64$

Factor each polynomial, if possible. If the polynomial cannot be factored, write *prime*.

4. $98x^2 - 200y^2$ **5.** $x^2 + 22x + 121$ **6.** $81 + 18j + j^2$

7. $25c^2 - 10c - 1$ **8.** $169 - 26r + r^2$ **9.** $7x^2 - 9x + 2$

10. $16m^2 + 48m + 36$ **11.** $16 - 25a^2$ **12.** $b^2 - 16b + 256$

13. $36x^2 - 12x + 1$ **14.** $16a^2 - 40ab + 25b^2$ **15.** $8m^3 - 64m$

8-9 Study Guide and Intervention (continued)

Perfect Squares

Solve Equations with Perfect Squares Factoring and the Zero Product Property can be used to solve equations that involve repeated factors. The repeated factor gives just one solution to the equation. You may also be able to use the **Square Root Property** below to solve certain equations.

Square Root Property	For any number $n > 0$, if $x^2 = n$, then $x = \pm\sqrt{n}$.

Example Solve each equation. Check your solutions.

a. $x^2 - 6x + 9 = 0$

$$x^2 - 6x + 9 = 0 \qquad \text{Original equation}$$
$$x^2 - 2(3x) + 3^2 = 0 \qquad \text{Recognize a perfect square trinomial.}$$
$$(x - 3)(x - 3) = 0 \qquad \text{Factor the perfect square trinomial.}$$
$$x - 3 = 0 \qquad \text{Set repeated factor equal to 0.}$$
$$x = 3 \qquad \text{Solve.}$$

The solution set is {3}. Since $3^2 - 6(3) + 9 = 0$, the solution checks.

b. $(a - 5)^2 = 64$

$$(a - 5)^2 = 64 \qquad \text{Original equation}$$
$$a - 5 = \pm\sqrt{64} \qquad \text{Square Root Property}$$
$$a - 5 = \pm 8 \qquad 64 = 8 \cdot 8$$
$$a = 5 \pm 8 \qquad \text{Add 5 to each side.}$$
$$a = 5 + 8 \quad \text{or} \quad a = 5 - 8 \qquad \text{Separate into 2 equations.}$$
$$a = 13 \qquad\qquad a = -3 \qquad \text{Solve each equation.}$$

The solution set is {-3, 13}. Since $(-3 - 5)^2 = 64$ and $(13 - 5)^2 = 64$, the solutions check.

Exercises

Solve each equation. Check the solutions.

1. $x^2 + 4x + 4 = 0$

2. $16n^2 + 16n + 4 = 0$

3. $25d^2 - 10d + 1 = 0$

4. $x^2 + 10x + 25 = 0$

5. $9x^2 - 6x + 1 = 0$

6. $x^2 + x + \dfrac{1}{4} = 0$

7. $25k^2 + 20k + 4 = 0$

8. $p^2 + 2p + 1 = 49$

9. $x^2 + 4x + 4 = 64$

10. $x^2 - 6x + 9 = 25$

11. $a^2 + 8a + 16 = 1$

12. $16y^2 + 8y + 1 = 0$

13. $(x + 3)^2 = 49$

14. $(y + 6)^2 = 1$

15. $(m - 7)^2 = 49$

16. $(2x + 1)^2 = 1$

17. $(4x + 3)^2 = 25$

18. $(3h - 2)^2 = 4$

19. $(x + 1)^2 = 7$

20. $(y - 3)^2 = 6$

21. $(m - 2)^2 = 5$

9-1 Study Guide and Intervention

Graphing Quadratic Functions

Characteristics of Quadratic Functions

Quadratic Function	a function described by an equation of the form $f(x) = ax^2 + bx + c$, where $a \neq 0$	Example: $y = 2x^2 + 3x + 8$

The parent graph of the family of quadratic functions is $y = x^2$. Graphs of quadratic functions have a general shape called a **parabola**. A parabola opens upward and has a **minimum point** when the value of a is positive, and a parabola opens downward and has a **maximum point** when the value of a is negative.

Example 1

a. **Use a table of values to graph** $y = x^2 - 4x + 1$.

x	y
−1	6
0	1
1	−2
2	−3
3	−2
4	1

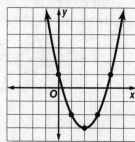

Graph the ordered pairs in the table and connect them with a smooth curve.

b. **What are the domain and range of this function?**

The domain is all real numbers. The range is all real numbers greater than or equal to −3, which is the minimum.

Example 2

a. **Use a table of values to graph** $y = -x^2 - 6x - 7$.

x	y
−6	−7
−5	−2
−4	1
−3	2
−2	1
−1	−2
0	−7

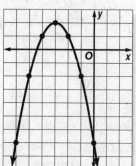

Graph the ordered pairs in the table and connect them with a smooth curve.

b. **What are the domain and range of this function?**

The domain is all real numbers. The range is all real numbers less than or equal to 2, which is the maximum.

Exercises

Use a table of values to graph each function. Determine the domain and range.

1. $y = x^2 + 2$

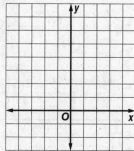

2. $y = -x^2 - 4$

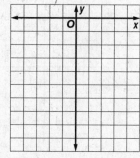

3. $y = x^2 - 3x + 2$

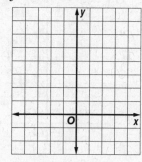

9-1 Study Guide and Intervention (continued)

Graphing Quadratic Functions

Symmetry and Vertices Parabolas have a geometric property called **symmetry**. That is, if the figure is folded in half, each half will match the other half exactly. The vertical line containing the fold line is called the **axis of symmetry**. The axis of symmetry contains the minimum or maximum point of the parabola, the **vertex**.

Axis of Symmetry	For the parabola $y = ax^2 + bx + c$, where $a \neq 0$, the line $x = -\dfrac{b}{2a}$ is the axis of symmetry.	**Example:** The axis of symmetry of $y = x^2 + 2x + 5$ is the line $x = -1$.

Example Consider the graph of $y = 2x^2 + 4x + 1$.

a. Write the equation of the axis of symmetry.

In $y = 2x^2 + 4x + 1$, $a = 2$ and $b = 4$. Substitute these values into the equation of the axis of symmetry.

$$x = -\frac{b}{2a}$$

$$x = -\frac{4}{2(2)} = -1$$

The axis of symmetry is $x = -1$.

b. Find the coordinates of the vertex.

Since the equation of the axis of symmetry is $x = -1$ and the vertex lies on the axis, the x-coordinate of the vertex is -1.

$y = 2x^2 + 4x + 1$	Original equation
$y = 2(-1)^2 + 4(-1) + 1$	Substitute.
$y = 2(1) - 4 + 1$	Simplify.
$y = -1$	

The vertex is at $(-1, -1)$.

c. Identify the vertex as a maximum or a minimum.

Since the coefficient of the x^2-term is positive, the parabola opens upward, and the vertex is a minimum point.

d. Graph the function.

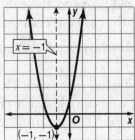

Exercises

Consider each equation. Determine whether the function has *maximum* or *minimum* value. State the maximum or minimum value and the domain and range of the function. Find the equation of the axis of symmetry. Graph the function.

1. $y = x^2 + 3$

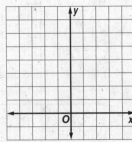

2. $y = -x^2 - 4x - 4$

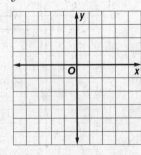

3. $y = x^2 + 2x + 3$

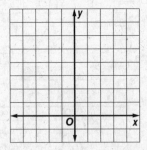

9-2 Study Guide and Intervention

Solving Quadratic Equations by Graphing

Solve by Graphing

Quadratic Equation	an equation of the form $ax^2 + bx + c = 0$, where $a \neq 0$

The solutions of a quadratic equation are called the **roots** of the equation. The roots of a quadratic equation can be found by graphing the related quadratic function $f(x) = ax^2 + bx + c$ and finding the x-intercepts or **zeros** of the function.

Example 1 Solve $x^2 + 4x + 3 = 0$ by graphing.

Graph the related function $f(x) = x^2 + 4x + 3$. The equation of the axis of symmetry is $x = -\dfrac{4}{2(1)}$ or -2. The vertex is at $(-2, -1)$. Graph the vertex and several other points on either side of the axis of symmetry.

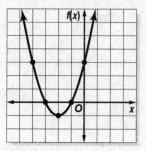

To solve $x^2 + 4x + 3 = 0$, you need to know where $f(x) = 0$. This occurs at the x-intercepts, -3 and -1.

The solutions are -3 and -1.

Example 2 Solve $x^2 - 6x + 9 = 0$ by graphing.

Graph the related function $f(x) = x^2 - 6x + 9$. The equation of the axis of symmetry is $x = \dfrac{6}{2(1)}$ or 3. The vertex is at $(3, 0)$. Graph the vertex and several other points on either side of the axis of symmetry.

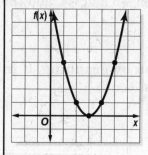

To solve $x^2 - 6x + 9 = 0$, you need to know where $f(x) = 0$. The vertex of the parabola is the x-intercept. Thus, the only solution is 3.

Exercises

Solve each equation by graphing.

1. $x^2 + 7x + 12 = 0$

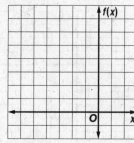

2. $x^2 - x - 12 = 0$

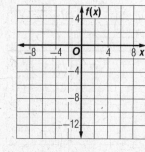

3. $x^2 - 4x + 5 = 0$

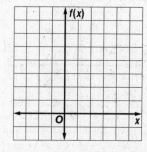

9-2 Study Guide and Intervention (continued)

Solving Quadratic Equations by Graphing

Estimate Solutions The roots of a quadratic equation may not be integers. If exact roots cannot be found, they can be estimated by finding the consecutive integers between which the roots lie.

Example Solve $x^2 + 6x + 6 = 0$ by graphing. If integral roots cannot be found, estimate the roots by stating the consecutive integers between which the roots lie.

Graph the related function $f(x) = x^2 + 6x + 6$.

x	f(x)
−5	1
−4	−2
−3	−3
−2	−2
−1	1

Notice that the value of the function changes from negative to positive between the x-values of −5 and −4 and between −2 and −1.

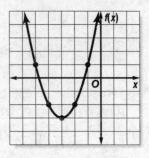

The x-intercepts of the graph are between −5 and −4 and between −2 and −1.
So one root is between −5 and −4, and the other root is between −2 and −1.

Exercises

Solve each equation by graphing. If integral roots cannot be found, estimate the roots to the nearest tenth.

1. $x^2 + 7x + 9 = 0$

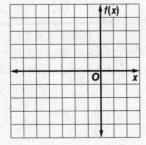

2. $x^2 - x - 4 = 0$

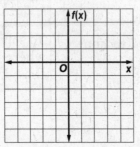

3. $x^2 - 4x + 6 = 0$

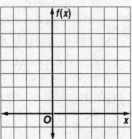

4. $x^2 - 4x - 1 = 0$

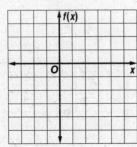

5. $4x^2 - 12x + 3 = 0$

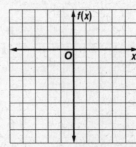

6. $x^2 - 2x - 4 = 0$

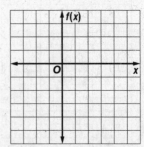

9-3 Study Guide and Intervention

Transformations of Quadratic Functions

Translations A **translation** is a change in the position of a figure either up, down, left, right, or diagonal. Adding or subtracting constants in the equations of functions translates the graphs of the functions.

The graph of $g(x) = x^2 + k$ translates the graph of $f(x) = x^2$ vertically.

If $k > 0$, the graph of $f(x) = x^2$ is translated k units up.

If $k < 0$, the graph of $f(x) = x^2$ is translated $|k|$ units down.

The graph of $g(x) = (x - h)^2$ is the graph of $f(x) = x^2$ translated horizontally.

If $h > 0$, the graph of $f(x) = x^2$ is translated h units to the right.

If $h < 0$, the graph of $f(x) = x^2$ is translated $|h|$ units to the left.

Example Describe how the graph of each function is related to the graph of $f(x) = x^2$.

a. $g(x) = x^2 + 4$

The value of k is 4, and $4 > 0$. Therefore, the graph of $g(x) = x^2 + 4$ is a translation of the graph of $f(x) = x^2$ up 4 units.

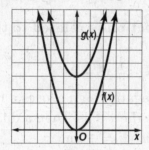

b. $g(x) = (x + 3)^2$

The value of h is -3, and $-3 < 0$. Thus, the graph of $g(x) = (x + 3)^2$ is a translation of the graph of $f(x) = x^2$ to the left 3 units.

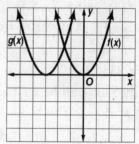

Exercises

Describe how the graph of each function is related to the graph of $f(x) = x^2$.

1. $g(x) = x^2 + 1$

2. $g(x) = (x - 6)^2$

3. $g(x) = (x + 1)^2$

4. $g(x) = 20 + x^2$

5. $g(x) = (-2 + x)^2$

6. $g(x) = -\dfrac{1}{2} + x^2$

7. $g(x) = x^2 + \dfrac{8}{9}$

8. $g(x) = x^2 - 0.3$

9. $g(x) = (x + 4)^2$

9-3   Study Guide and Intervention (continued)

Transformations of Quadratic Functions

Dilations and Reflections A **dilation** is a transformation that makes the graph narrower or wider than the parent graph. A **reflection** flips a figure over the *x*- or *y*-axis.

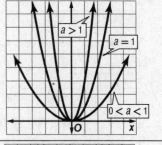

The graph of $f(x) = ax^2$ stretches or compresses the graph of $f(x) = x^2$.

If $|a| > 1$, the graph of $f(x) = x^2$ is stretched vertically.

If $0 < |a| < 1$, the graph of $f(x) = x^2$ is compressed vertically.

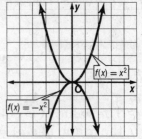

The graph of the function $-f(x)$ flips the graph of $f(x) = x^2$ across the *x*-axis.

The graph of the function $f(-x)$ flips the graph of $f(x) = x^2$ across the *y*-axis.

Example   Describe how the graph of each function is related to the graph of $f(x) = x^2$.

a. $g(x) = 2x^2$

The function can be written as $f(x) = ax^2$ where $a = 2$. Because $|a| > 1$, the graph of $y = 2x^2$ is the graph of $y = x^2$ that is stretched vertically.

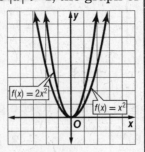

b. $g(x) = -\dfrac{1}{2}x^2 - 3$

The negative sign causes a reflection across the *x*-axis. Then a dilation occurs in which $a = \dfrac{1}{2}$ and a translation in which $k = -3$. So the graph of $g(x) = -\dfrac{1}{2}x^2 - 3$ is reflected across the *x*-axis, dilated wider than the graph of $f(x) = x^2$, and translated down 3 units.

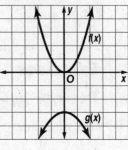

Exercises

Describe how the graph of each function is related to the graph of $f(x) = x^2$.

1. $g(x) = -5x^2$

2. $g(x) = -(x + 1)^2$

3. $g(x) = -\dfrac{1}{4}x^2 - 1$

9-4 Study Guide and Intervention

Solving Quadratic Equations by Completing the Square

Complete the Square Perfect square trinomials can be solved quickly by taking the square root of both sides of the equation. A quadratic equation that is not in perfect square form can be made into a perfect square by a method called **completing the square**.

Completing the Square

To complete the square for any quadratic equation of the form $x^2 + bx$:

Step 1 Find one-half of b, the coefficient of x.

Step 2 Square the result in Step 1.

Step 3 Add the result of Step 2 to $x^2 + bx$.

$$x^2 + bx + \left(\frac{b}{2}\right)^2 = \left(x + \frac{b}{2}\right)^2$$

Example **Find the value of c that makes $x^2 + 2x + c$ a perfect square trinomial.**

Step 1 Find $\frac{1}{2}$ of 2. $\frac{2}{2} = 1$

Step 2 Square the result of Step 1. $1^2 = 1$

Step 3 Add the result of Step 2 to $x^2 + 2x$. $x^2 + 2x + 1$

Thus, $c = 1$. Notice that $x^2 + 2x + 1$ equals $(x + 1)^2$.

Exercises

Find the value of c that makes each trinomial a perfect square.

1. $x^2 + 10x + c$ 2. $x^2 + 14x + c$

3. $x^2 - 4x + c$ 4. $x^2 - 8x + c$

5. $x^2 + 5x + c$ 6. $x^2 + 9x + c$

7. $x^2 - 3x + c$ 8. $x^2 - 15x + c$

9. $x^2 + 28x + c$ 10. $x^2 + 22x + c$

9-4 Study Guide and Intervention (continued)

Solving Quadratic Equations by Completing the Square

Solve by Completing the Square Since few quadratic expressions are perfect square trinomials, the method of **completing the square** can be used to solve some quadratic equations. Use the following steps to complete the square for a quadratic expression of the form $ax^2 + bx$.

Step 1	Find $\frac{b}{2}$.
Step 2	Find $\left(\frac{b}{2}\right)^2$.
Step 3	Add $\left(\frac{b}{2}\right)^2$ to $ax^2 + bx$.

Example **Solve $x^2 + 6x + 3 = 10$ by completing the square.**

$x^2 + 6x + 3 = 10$	Original equation
$x^2 + 6x + 3 - 3 = 10 - 3$	Subtract 3 from each side.
$x^2 + 6x = 7$	Simplify.
$x^2 + 6x + 9 = 7 + 9$	Since $\left(\frac{6}{2}\right)^2 = 9$, add 9 to each side.
$(x + 3)^2 = 16$	Factor $x^2 + 6x + 9$.
$x + 3 = \pm 4$	Take the square root of each side.
$x = -3 \pm 4$	Simplify.

$x = -3 + 4$ or $x = -3 - 4$
$\quad = 1 \qquad\qquad = -7$

The solution set is $\{-7, 1\}$.

Exercises

Solve each equation by completing the square. Round to the nearest tenth if necessary.

1. $x^2 - 4x + 3 = 0$ **2.** $x^2 + 10x = -9$ **3.** $x^2 - 8x - 9 = 0$

4. $x^2 - 6x = 16$ **5.** $x^2 - 4x - 5 = 0$ **6.** $x^2 - 12x = 9$

7. $x^2 + 8x = 20$ **8.** $x^2 = 2x + 1$ **9.** $x^2 + 20x + 11 = -8$

10. $x^2 - 1 = 5x$ **11.** $x^2 = 22x + 23$ **12.** $x^2 - 8x = -7$

13. $x^2 + 10x = 24$ **14.** $x^2 - 18x = 19$ **15.** $x^2 + 16x = -16$

16. $4x^2 = 24 + 4x$ **17.** $2x^2 + 4x + 2 = 8$ **18.** $4x^2 = 40x + 44$

9-5 Study Guide and Intervention

Solving Quadratic Equations by Using the Quadratic Formula

Quadratic Formula To solve the standard form of the quadratic equation, $ax^2 + bx + c = 0$, use the **Quadratic Formula**.

Quadratic Formula	The solutions of $ax^2 + bx + c = 0$, where $a \neq 0$, are given by $x = \dfrac{-b \pm \sqrt{b^2 - 4ac}}{2a}$.

Example 1 **Solve $x^2 + 2x = 3$ by using the Quadratic Formula.**

Rewrite the equation in standard form.

$$x^2 + 2x = 3 \qquad \text{Original equation}$$
$$x^2 + 2x - 3 = 3 - 3 \qquad \text{Subtract 3 from each side.}$$
$$x^2 + 2x - 3 = 0 \qquad \text{Simplify.}$$

Now let $a = 1$, $b = 2$, and $c = -3$ in the Quadratic Formula.

$$x = \frac{-b \pm \sqrt{b^2 - 4ac}}{2a}$$

$$= \frac{-2 \pm \sqrt{(2)^2 - 4(1)(-3)}}{2(1)}$$

$$= \frac{-2 \pm \sqrt{16}}{2}$$

$$x = \frac{-2 + 4}{2} \quad \text{or} \quad x = \frac{-2 - 4}{2}$$
$$= 1 \qquad\qquad = -3$$

The solution set is $\{-3, 1\}$.

Example 2 **Solve $x^2 - 6x - 2 = 0$ by using the Quadratic Formula. Round to the nearest tenth if necessary.**

For this equation $a = 1$, $b = -6$, and $c = -2$.

$$x = \frac{-b \pm \sqrt{b^2 - 4ac}}{2a}$$

$$= \frac{6 \pm \sqrt{(-6)^2 - 4(1)(-2)}}{2(1)}$$

$$= \frac{6 + \sqrt{44}}{2}$$

$$x = \frac{6 + \sqrt{44}}{2} \quad \text{or} \quad x = \frac{6 - \sqrt{44}}{2}$$
$$\approx 6.3 \qquad\qquad\qquad \approx -0.3$$

The solution set is $\{-0.3, 6.3\}$.

Exercises

Solve each equation by using the Quadratic Formula. Round to the nearest tenth if necessary.

1. $x^2 - 3x + 2 = 0$

2. $x^2 - 8x = -16$

3. $16x^2 - 8x = -1$

4. $x^2 + 5x = 6$

5. $3x^2 + 2x = 8$

6. $8x^2 - 8x - 5 = 0$

7. $-4x^2 + 19x = 21$

8. $2x^2 + 6x = 5$

9. $48x^2 + 22x - 15 = 0$

10. $8x^2 - 4x = 24$

11. $2x^2 + 5x = 8$

12. $8x^2 + 9x - 4 = 0$

13. $2x^2 + 9x + 4 = 0$

14. $8x^2 + 17x + 2 = 0$

9-5 Study Guide and Intervention (continued)

Solving Quadratic Equations by Using the Quadratic Formula

The Discriminant In the Quadratic Formula, $x = \dfrac{-b \pm \sqrt{b^2 - 4ac}}{2a}$, the expression under the radical sign, $b^2 - 4ac$, is called the **discriminant**. The discriminant can be used to determine the number of real solutions for a quadratic equation.

Case 1: $b^2 - 4ac < 0$	no real solutions
Case 2: $b^2 - 4ac = 0$	one real solution
Case 3: $b^2 - 4ac > 0$	two real solutions

Example State the value of the discriminant for each equation. Then determine the number of real solutions of the equation.

a. $12x^2 + 5x = 4$

Write the equation in standard form.

$12x^2 + 5x = 4$	Original equation
$12x^2 + 5x - 4 = 4 - 4$	Subtract 4 from each side.
$12x^2 + 5x - 4 = 0$	Simplify.

Now find the discriminant.

$b^2 - 4ac = (5)^2 - 4(12)(-4)$

$\qquad\qquad = 217$

Since the discriminant is positive, the equation has two real solutions.

b. $2x^2 + 3x = -4$

$2x^2 + 3x = -4$	Original equation
$2x^2 + 3x + 4 = -4 + 4$	Add 4 to each side.
$2x^2 + 3x + 4 = 0$	Simplify.

Find the discriminant.

$b^2 - 4ac = (3)^2 - 4(2)(4)$

$\qquad\qquad = -23$

Since the discriminant is negative, the equation has no real solutions.

Exercises

State the value of the discriminant for each equation. Then determine the number of real solutions of the equation.

1. $3x^2 + 2x - 3 = 0$

2. $3x^2 - 7x - 8 = 0$

3. $2x^2 - 10x - 9 = 0$

4. $4x^2 = x + 4$

5. $3x^2 - 13x = 10$

6. $6x^2 - 10x + 10 = 0$

7. $2x^2 - 20 = -x$

8. $6x^2 = -11x - 40$

9. $9 - 18x + 9x^2 = 0$

10. $12x^2 + 9 = -6x$

11. $9x^2 = 81$

12. $16x^2 + 16x + 4 = 0$

13. $8x^2 + 9x = 2$

14. $4x^2 - 4x + 4 = 3$

15. $3x^2 - 18x = -14$

9-6 Study Guide and Intervention

Analyzing Functions with Successive Differences and Ratios

Identify Functions Linear functions, quadratic functions, and exponential functions can all be used to model data. The general forms of the equations are listed at the right.

Linear Function	$y = mx + b$
Quadratic Function	$y = ax^2 + bx + c$
Exponential Function	$y = ab^x$

You can also identify data as linear, quadratic, or exponential based on patterns of behavior of their y-values.

Example 1 Graph the set of ordered pairs {(–3, 2), (–2, –1), (–1, –2), (0, –1), (1, 2)}. Determine whether the ordered pairs represent a *linear* function, a *quadratic* function, or an *exponential* function.

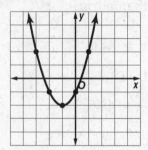

The ordered pairs appear to represent a quadratic function.

Example 2 Look for a pattern in the table to determine which model best describes the data.

x	–2	–1	0	1	2
y	4	2	1	0.5	0.25

Start by comparing the first differences.

$$4 \xrightarrow{-2} 2 \xrightarrow{-1} 1 \xrightarrow{-0.5} 0.5 \xrightarrow{-0.25} 0.25$$

The first differences are not all equal. The table does not represent a linear function. Find the second differences and compare.

$$-2 \xrightarrow{+1} -1 \xrightarrow{+0.5} -0.5 \xrightarrow{+0.25} -0.25$$

The table does not represent a quadratic function. Find the ratios of the y-values.

$$4 \xrightarrow{\times 0.5} 2 \xrightarrow{\times 0.5} 1 \xrightarrow{\times 0.5} 0.5 \xrightarrow{\times 0.5} 0.25$$

The ratios are equal. Therefore, the table can be modeled by an exponential function.

Exercises

Graph each set of ordered pairs. Determine whether the ordered pairs represent a *linear* function, a *quadratic* function, or an *exponential* function.

1. (0, –1), (1, 1), (2, 3), (3, 5)

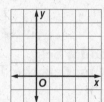

2. (–3, –1), (–2, –4), (–1, –5), (0, –4), (1, –1)

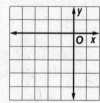

Look for a pattern in each table to determine which model best describes the data.

3.

x	–2	–1	0	1	2
y	6	5	4	3	2

4.

x	–2	–1	0	1	2
y	6.25	2.5	1	0.4	0.16

9-6 Study Guide and Intervention (continued)

Analyzing Functions with Successive Differences and Ratios

Write Equations Once you find the model that best describes the data, you can write an equation for the function.

Basic Forms	Linear Function	$y = mx + b$
	Quadratic Function	$y = ax^2$
	Exponential Function	$y = ab^x$

Example Determine which model best describes the data. Then write an equation for the function that models the data.

x	0	1	2	3	4
y	3	6	12	24	48

Step 1 Determine whether the data is modeled by a linear, quadratic, or exponential function.

First differences: $3 \xrightarrow{+3} 6 \xrightarrow{+6} 12 \xrightarrow{+12} 24 \xrightarrow{+24} 48$

Second differences: $3 \xrightarrow{+3} 6 \xrightarrow{+6} 12 \xrightarrow{+12} 24$

y-value ratios: $3 \xrightarrow{\times 2} 6 \xrightarrow{\times 2} 12 \xrightarrow{\times 2} 24 \xrightarrow{\times 2} 48$

The ratios of successive y-values are equal. Therefore, the table of values can be modeled by an exponential function.

Step 2 Write an equation for the function that models the data. The equation has the form $y = ab^x$. The y-value ratio is 2, so this is the value of the base.

$y = ab^x$ Equation for exponential function

$3 = a(2)^0$ $x = 0, y = 3$, and $b = 2$

$3 = a$ Simplify.

An equation that models the data is $y = 3 \cdot 2^x$. To check the results, you can verify that the other ordered pairs satisfy the function.

Exercises

Look for a pattern in each table of values to determine which model best describes the data. Then write an equation for the function that models the data.

1.
x	−2	−1	0	1	2
y	12	3	0	3	12

2.
x	−1	0	1	2	3
y	−2	1	4	7	10

3.
x	−1	0	1	2	3
y	0.75	3	12	48	192

9-7 Study Guide and Intervention

Special Functions

Step Functions The graph of a **step function** is a series of disjointed line segments. Because each part of a step function is linear, this type of function is called a **piecewise-linear function**.

One example of a step function is the greatest integer function, written as $f(x) = [\![x]\!]$, where $f(x)$ is the greatest integer not greater than x.

Example **Graph $f(x) = [\![x + 3]\!]$.**

Make a table of values using integer and noninteger values. On the graph, dots represent included points, and circles represent points that are excluded.

x	$x + 3$	$[\![x + 3]\!]$
−5	−2	−2
−3.5	−0.5	−1
−2	1	1
−0.5	2.5	2
1	4	4
2.5	5.5	5

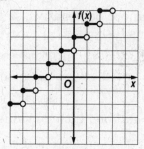

Because the dots and circles overlap, the domain is all real numbers. The range is all integers.

Exercises

Graph each function. State the domain and range.

1. $f(x) = [\![x + 1]\!]$

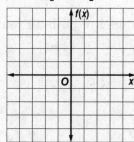

2. $f(x) = -[\![x]\!]$

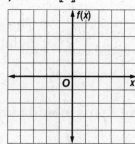

3. $f(x) = [\![x - 1]\!]$

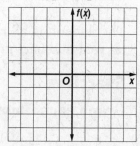

4. $f(x) = [\![x]\!] + 4$

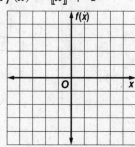

5. $f(x) = [\![x]\!] - 3$

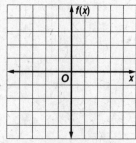

6. $f(x) = [\![2x]\!]$

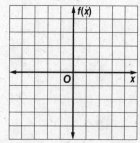

9-7   **Study Guide and Intervention** *(continued)*
Special Functions

Absolute Value Functions Another type of piecewise-linear function is the **absolute value function**. Recall that the absolute value of a number is always nonnegative. So in the absolute value function, written as $f(x) = |x|$, all of the values of the range are nonnegative.

The absolute value function is called a **piecewise-defined function** because it can be written using two or more expressions.

Example 1   Graph $f(x) = |x + 2|$. **State the domain and range.**

$f(x)$ cannot be negative, so the minimum point is $f(x) = 0$.

$f(x) = |x + 2|$	Original function
$0 = x + 2$	Replace $f(x)$ with 0.
$-2 = x$	Subtract 2 from each side.

Make a table. Include values for $x > -2$ and $x < -2$.

x	f(x)
−5	3
−4	2
−3	1
−2	0
−1	1
0	2
1	3
2	4

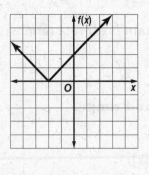

The domain is all real numbers. The range is all real numbers greater than or equal to 0.

Example 2   Graph
$$f(x) = \begin{cases} x + 1 \text{ if } x > 1 \\ 3x \text{ if } x \le 1 \end{cases}.$$ **State the domain and range.**

Graph the first expression. When $x > 1$, $f(x) = x + 1$. Since $x \ne 1$, place an open circle at (1, 2).

Next, graph the second expression. When $x \le 1$, $f(x) = 3x$. Since $x = 1$, place a closed circle at (1, 3).

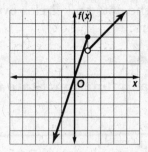

The domain and range are both all real numbers.

Exercises

Graph each function. State the domain and range.

1. $f(x) = |x - 1|$

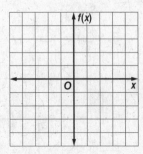

2. $f(x) = |-x + 2|$

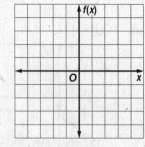

3. $f(x) = \begin{cases} -x + 4 \text{ if } x \le 1 \\ x - 2 \text{ if } x > 1 \end{cases}$

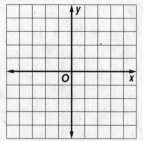

10-1 Study Guide and Intervention

Square Root Functions

Dilations of Radical Functions A **square root function** contains the square root of a variable. Square root functions are a type of **radical function**.

In order for a square root to be a real number, the **radicand**, or the expression under the radical sign, cannot be negative. Values that make the radicand negative are not included in the domain.

Square Root Function	Parent function: $f(x) = \sqrt{x}$ Type of graph: curve Domain: $\{x \mid x \geq 0\}$ Range: $\{y \mid y \geq 0\}$

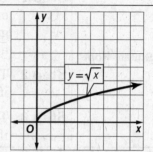

Example **Graph $y = 3\sqrt{x}$. State the domain and range.**

Step 1 Make a table. Choose nonnegative values for x.

x	y
0	0
0.5	≈ 2.12
1	3
2	≈ 4.24
4	6
6	≈ 7.35

Step 2 Plot points and draw a smooth curve.

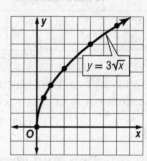

The domain is $\{x \mid x \geq 0\}$ and the range is $\{y \mid y \geq 0\}$.

Exercises

Graph each function, and compare to the parent graph. State the domain and range.

1. $y = \dfrac{3}{2}\sqrt{x}$

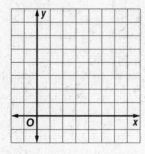

2. $y = 4\sqrt{x}$

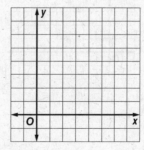

3. $y = \dfrac{5}{2}\sqrt{x}$

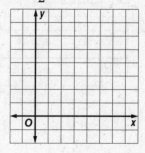

10-1 **Study Guide and Intervention** (continued)

Square Root Functions

Reflections and Translations of Radical Functions Radical functions, like quadratic functions, can be translated horizontally and vertically, as well as reflected across the x-axis. To draw the graph of $y = a\sqrt{x + h} + k$, follow these steps.

Graphs of Square Root Functions	**Step 1**	Draw the graph of $y = a\sqrt{x}$. The graph starts at the origin and passes through the point at $(1, a)$. If $a > 0$, the graph is in the 1st quadrant. If $a < 0$, the graph is reflected across the x-axis and is in the 4th quadrant.
	Step 2	Translate the graph $|k|$ units up if k is positive and down if k is negative.
	Step 3	Translate the graph $|h|$ units left if h is positive and right if h is negative.

Example Graph $y = -\sqrt{x + 1}$ and compare to the parent graph. State the domain and range.

Step 1 Make a table of values.

x	−1	0	1	3	8
y	0	−1	−1.41	−2	−3

Step 2 This is a horizontal translation 1 unit to the left of the parent function and reflected across the x-axis. The domain is $\{x | x \geq -1\}$ and the range is $\{y | y \leq 0\}$.

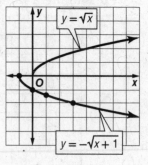

Exercises

Graph each function, and compare to the parent graph. State the domain and range.

1. $y = \sqrt{x} + 3$

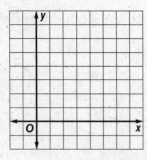

2. $y = \sqrt{x - 1}$

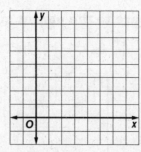

3. $y = -\sqrt{x - 1}$

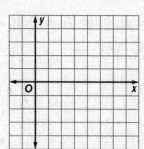

10-2 Study Guide and Intervention

Simplifying Radical Expressions

Product Property of Square Roots The **Product Property of Square Roots** and prime factorization can be used to simplify expressions involving irrational square roots. When you simplify radical expressions with variables, use absolute value to ensure nonnegative results.

Product Property of Square Roots	For any numbers a and b, where $a \geq 0$ and $b \geq 0$, $\sqrt{ab} = \sqrt{a} \cdot \sqrt{b}$.

Example 1 Simplify $\sqrt{180}$.

$$\sqrt{180} = \sqrt{2 \cdot 2 \cdot 3 \cdot 3 \cdot 5} \qquad \text{Prime factorization of 180}$$
$$= \sqrt{2^2} \cdot \sqrt{3^2} \cdot \sqrt{5} \qquad \text{Product Property of Square Roots}$$
$$= 2 \cdot 3 \cdot \sqrt{5} \qquad \text{Simplify.}$$
$$= 6\sqrt{5} \qquad \text{Simplify.}$$

Example 2 Simplify $\sqrt{120a^2 \cdot b^5 \cdot c^4}$.

$$\sqrt{120a^2 \cdot b^5 \cdot c^4}$$
$$= \sqrt{2^3 \cdot 3 \cdot 5 \cdot a^2 \cdot b^5 \cdot c^4}$$
$$= \sqrt{2^2} \cdot \sqrt{2} \cdot \sqrt{3} \cdot \sqrt{5} \cdot \sqrt{a^2} \cdot \sqrt{b^4 \cdot b} \cdot \sqrt{c^4}$$
$$= 2 \cdot \sqrt{2} \cdot \sqrt{3} \cdot \sqrt{5} \cdot |a| \cdot b^2 \cdot \sqrt{b} \cdot c^2$$
$$= 2|a|b^2c^2\sqrt{30b}$$

Exercises

Simplify each expression.

1. $\sqrt{28}$

2. $\sqrt{68}$

3. $\sqrt{60}$

4. $\sqrt{75}$

5. $\sqrt{162}$

6. $\sqrt{3} \cdot \sqrt{6}$

7. $\sqrt{2} \cdot \sqrt{5}$

8. $\sqrt{5} \cdot \sqrt{10}$

9. $\sqrt{4a^2}$

10. $\sqrt{9x^4}$

11. $\sqrt{300a^4}$

12. $\sqrt{128c^6}$

13. $4\sqrt{10} \cdot 3\sqrt{6}$

14. $\sqrt{3x^2} \cdot 3\sqrt{3x^4}$

15. $\sqrt{20a^2b^4}$

16. $\sqrt{100x^3y}$

17. $\sqrt{24a^4b^2}$

18. $\sqrt{81x^4y^2}$

19. $\sqrt{150a^2b^2c}$

20. $\sqrt{72a^6b^3c^2}$

21. $\sqrt{45x^2y^5z^8}$

22. $\sqrt{98x^4y^6z^2}$

10-2 **Study Guide and Intervention** (continued)

Simplifying Radical Expressions

Quotient Property of Square Roots A fraction containing radicals is in simplest form if no radicals are left in the denominator. The **Quotient Property of Square Roots** and **rationalizing the denominator** can be used to simplify radical expressions that involve division. When you rationalize the denominator, you multiply the numerator and denominator by a radical expression that gives a rational number in the denominator.

Quotient Property of Square Roots	For any numbers a and b, where $a \geq 0$ and $b > 0$, $\sqrt{\dfrac{a}{b}} = \dfrac{\sqrt{a}}{\sqrt{b}}$.

Example Simplify $\sqrt{\dfrac{56}{45}}$.

$\sqrt{\dfrac{56}{45}} = \sqrt{\dfrac{4 \cdot 14}{9 \cdot 5}}$ Factor 56 and 45.

$= \dfrac{2 \cdot \sqrt{14}}{3 \cdot \sqrt{5}}$ Simplify the numerator and denominator.

$= \dfrac{2\sqrt{14}}{3\sqrt{5}} \cdot \dfrac{\sqrt{5}}{\sqrt{5}}$ Multiply by $\dfrac{\sqrt{5}}{\sqrt{5}}$ to rationalize the denominator.

$= \dfrac{2\sqrt{70}}{15}$ Product Property of Square Roots

Exercises

Simplify each expression.

1. $\dfrac{\sqrt{9}}{\sqrt{18}}$

2. $\dfrac{\sqrt{8}}{\sqrt{24}}$

3. $\dfrac{\sqrt{100}}{\sqrt{121}}$

4. $\dfrac{\sqrt{75}}{\sqrt{3}}$

5. $\dfrac{8\sqrt{2}}{2\sqrt{8}}$

6. $\sqrt{\dfrac{2}{5}} \cdot \sqrt{\dfrac{6}{5}}$

7. $\sqrt{\dfrac{3}{4}} \cdot \sqrt{\dfrac{5}{2}}$

8. $\sqrt{\dfrac{5}{7}} \cdot \sqrt{\dfrac{2}{5}}$

9. $\sqrt{\dfrac{3a^2}{10b^6}}$

10. $\sqrt{\dfrac{x^6}{y^4}}$

11. $\sqrt{\dfrac{100a^4}{144b^8}}$

12. $\sqrt{\dfrac{75b^3c^6}{a^2}}$

13. $\dfrac{\sqrt{4}}{3 - \sqrt{5}}$

14. $\dfrac{\sqrt{8}}{2 + \sqrt{3}}$

15. $\dfrac{\sqrt{5}}{5 + \sqrt{5}}$

16. $\dfrac{\sqrt{8}}{2\sqrt{7} + 4\sqrt{10}}$

10-3 Study Guide and Intervention

Operations with Radical Expressions

Add or Subtract Radical Expressions When adding or subtracting radical expressions, use the Associative and Distributive Properties to simplify the expressions. If radical expressions are not in simplest form, simplify them.

Example 1 Simplify $10\sqrt{6} - 5\sqrt{3} + 6\sqrt{3} - 4\sqrt{6}$.

$10\sqrt{6} - 5\sqrt{3} + 6\sqrt{3} - 4\sqrt{6} = (10 - 4)\sqrt{6} + (-5 + 6)\sqrt{3}$ Associative and Distributive Properties

$\qquad\qquad\qquad\qquad\qquad\quad = 6\sqrt{6} + \sqrt{3}$ Simplify.

Example 2 Simplify $3\sqrt{12} + 5\sqrt{75}$.

$3\sqrt{12} + 5\sqrt{75} = 3\sqrt{2^2 \cdot 3} + 5\sqrt{5^2 \cdot 3}$ Factor 12 and 75.

$\qquad\qquad\quad = 3 \cdot 2\sqrt{3} + 5 \cdot 5\sqrt{3}$ Simplify.

$\qquad\qquad\quad = 6\sqrt{3} + 25\sqrt{3}$ Multiply.

$\qquad\qquad\quad = 31\sqrt{3}$ Distributive Property

Exercises

Simplify each expression.

1. $2\sqrt{5} + 4\sqrt{5}$

2. $\sqrt{6} - 4\sqrt{6}$

3. $\sqrt{8} - \sqrt{2}$

4. $3\sqrt{75} + 2\sqrt{5}$

5. $\sqrt{20} + 2\sqrt{5} - 3\sqrt{5}$

6. $2\sqrt{3} + \sqrt{6} - 5\sqrt{3}$

7. $\sqrt{12} + 2\sqrt{3} - 5\sqrt{3}$

8. $3\sqrt{6} + 3\sqrt{2} - \sqrt{50} + \sqrt{24}$

9. $\sqrt{8a} - \sqrt{2a} + 5\sqrt{2a}$

10. $\sqrt{54} + \sqrt{24}$

11. $\sqrt{3} + \sqrt{\dfrac{1}{3}}$

12. $\sqrt{12} + \sqrt{\dfrac{1}{3}}$

13. $\sqrt{54} - \sqrt{\dfrac{1}{6}}$

14. $\sqrt{80} - \sqrt{20} + \sqrt{180}$

15. $\sqrt{50} + \sqrt{18} - \sqrt{75} + \sqrt{27}$

16. $2\sqrt{3} - 4\sqrt{45} + 2\sqrt{\dfrac{1}{3}}$

17. $\sqrt{125} - 2\sqrt{\dfrac{1}{5}} + \sqrt{\dfrac{1}{3}}$

18. $\sqrt{\dfrac{2}{3}} + 3\sqrt{3} - 4\sqrt{\dfrac{1}{12}}$

10-3 Study Guide and Intervention (continued)

Operations with Radical Expressions

Multiply Radical Expressions Multiplying two radical expressions with different radicands is similar to multiplying binomials.

Example Multiply $(3\sqrt{2} - 2\sqrt{5})(4\sqrt{20} + \sqrt{8})$.

Use the FOIL method.

$(3\sqrt{2} - 2\sqrt{5})(4\sqrt{20} + \sqrt{8}) = (3\sqrt{2})(4\sqrt{20}) + (3\sqrt{2})(\sqrt{8}) + (-2\sqrt{5})(4\sqrt{20}) + (-2\sqrt{5})(\sqrt{8})$

$= 12\sqrt{40} + 3\sqrt{16} - 8\sqrt{100} - 2\sqrt{40}$ Multiply.

$= 12\sqrt{2^2 \cdot 10} + 3 \cdot 4 - 8 \cdot 10 - 2\sqrt{2^2 \cdot 10}$ Simplify.

$= 24\sqrt{10} + 12 - 80 - 4\sqrt{10}$ Simplify.

$= 20\sqrt{10} - 68$ Combine like terms.

Exercises

Simplify each expression.

1. $2(\sqrt{3} + 4\sqrt{5})$

2. $\sqrt{6}(\sqrt{3} - 2\sqrt{6})$

3. $\sqrt{5}(\sqrt{5} - \sqrt{2})$

4. $\sqrt{2}(3\sqrt{7} + 2\sqrt{5})$

5. $(2 - 4\sqrt{2})(2 + 4\sqrt{2})$

6. $(3 + \sqrt{6})^2$

7. $(2 - 2\sqrt{5})^2$

8. $3\sqrt{2}(\sqrt{8} + \sqrt{24})$

9. $\sqrt{8}(\sqrt{2} + 5\sqrt{8})$

10. $(\sqrt{5} - 3\sqrt{2})(\sqrt{5} + 3\sqrt{2})$

11. $(\sqrt{3} + \sqrt{6})^2$

12. $(\sqrt{2} - 2\sqrt{3})^2$

13. $(\sqrt{5} - \sqrt{2})(\sqrt{2} + \sqrt{6})$

14. $(\sqrt{8} - \sqrt{2})(\sqrt{3} + \sqrt{6})$

15. $(\sqrt{5} - \sqrt{18})(7\sqrt{5} + \sqrt{3})$

16. $(2\sqrt{3} - \sqrt{45})(\sqrt{12} + 2\sqrt{6})$

17. $(2\sqrt{5} - 2\sqrt{3})(\sqrt{10} + \sqrt{6})$

18. $(\sqrt{2} + 3\sqrt{3})(\sqrt{12} - 4\sqrt{8})$

10-4 Study Guide and Intervention

Radical Equations

Radical Equations Equations containing radicals with variables in the radicand are called **radical equations**. These can be solved by first using the following steps.

Step 1 Isolate the radical on one side of the equation.
Step 2 Square each side of the equation to eliminate the radical.

Example 1 Solve $16 = \dfrac{\sqrt{x}}{2}$ for x.

$16 = \dfrac{\sqrt{x}}{2}$ Original equation

$2(16) = 2\left(\dfrac{\sqrt{x}}{2}\right)$ Multiply each side by 2.

$32 = \sqrt{x}$ Simplify.

$(32)^2 = (\sqrt{x})^2$ Square each side.

$1024 = x$ Simplify.

The solution is 1024, which checks in the original equation.

Example 2 Solve $\sqrt{4x - 7} + 2 = 7$.

$\sqrt{4x - 7} + 2 = 7$ Original equation

$\sqrt{4x - 7} + 2 - 2 = 7 - 2$ Subtract 2 from each side.

$\sqrt{4x - 7} = 5$ Simplify.

$(\sqrt{4x - 7})^2 = 5^2$ Square each side.

$4x - 7 = 25$ Simplify.

$4x - 7 + 7 = 25 + 7$ Add 7 to each side.

$4x = 32$ Simplify.

$x = 8$ Divide each side by 4.

The solution is 8, which checks in the original equation.

Exercises

Solve each equation. Check your solution.

1. $\sqrt{a} = 8$

2. $\sqrt{a} + 6 = 32$

3. $2\sqrt{x} = 8$

4. $7 = \sqrt{26 - n}$

5. $\sqrt{-a} = 6$

6. $\sqrt{3r^2} = 3$

7. $2\sqrt{3} = \sqrt{y}$

8. $2\sqrt{3a} - 2 = 7$

9. $\sqrt{x - 4} = 6$

10. $\sqrt{2m + 3} = 5$

11. $\sqrt{3b - 2} + 19 = 24$

12. $\sqrt{4x - 1} = 3$

13. $\sqrt{3r + 2} = 2\sqrt{3}$

14. $\sqrt{\dfrac{x}{2}} = \dfrac{1}{2}$

15. $\sqrt{\dfrac{x}{8}} = 4$

16. $\sqrt{6x^2 + 5x} = 2$

17. $\sqrt{\dfrac{x}{3}} + 6 = 8$

18. $2\sqrt{\dfrac{3x}{5}} + 3 = 11$

10-4 Study Guide and Intervention (continued)

Radical Equations

Extraneous Solutions To solve a radical equation with a variable on both sides, you need to square each side of the equation. Squaring each side of an equation sometimes produces **extraneous solutions**, or solutions that are not solutions of the original equation. Therefore, it is very important that you check each solution.

Example 1 Solve $\sqrt{x + 3} = x - 3$.

$$\sqrt{x + 3} = x - 3 \qquad \text{Original equation}$$
$$\left(\sqrt{x + 3}\right)^2 = (x - 3)^2 \qquad \text{Square each side.}$$
$$x + 3 = x^2 - 6x + 9 \qquad \text{Simplify.}$$
$$0 = x^2 - 7x + 6 \qquad \text{Subtract } x \text{ and 3 from each side.}$$
$$0 = (x - 1)(x - 6) \qquad \text{Factor.}$$
$$x - 1 = 0 \quad \text{or} \quad x - 6 = 0 \qquad \text{Zero Product Property}$$
$$x = 1 \qquad\qquad x = 6 \qquad \text{Solve.}$$

CHECK $\sqrt{x + 3} = x - 3$ $\qquad\qquad$ $\sqrt{x + 3} = x - 3$

$\sqrt{1 + 3} \overset{?}{=} 1 - 3$ $\qquad\qquad$ $\sqrt{6 + 3} \overset{?}{=} 6 - 3$

$\sqrt{4} \overset{?}{=} -2$ $\qquad\qquad\qquad$ $\sqrt{9} \overset{?}{=} 3$

$2 \neq -2$ $\qquad\qquad\qquad\qquad$ $3 = 3$ ✓

Since $x = 1$ does not satisfy the original equation, $x = 6$ is the only solution.

Exercises

Solve each equation. Check your solution.

1. $\sqrt{a} = a$

2. $\sqrt{a + 6} = a$

3. $2\sqrt{x} = x$

4. $n = \sqrt{2 - n}$

5. $\sqrt{-a} = a$

6. $\sqrt{10 - 6k} + 3 = k$

7. $\sqrt{y - 1} = y - 1$

8. $\sqrt{3a - 2} = a$

9. $\sqrt{x + 2} = x$

10. $\sqrt{2b + 5} = b - 5$

11. $\sqrt{3b + 6} = b + 2$

12. $\sqrt{4x - 4} = x$

13. $r + \sqrt{2 - r} = 2$

14. $\sqrt{x^2 + 10x} = x + 4$

15. $-2\sqrt{\dfrac{x}{8}} = 15$

16. $\sqrt{6x^2 - 4x} = x + 2$

17. $\sqrt{2y^2 - 64} = y$

18. $\sqrt{3x^2 + 12x + 1} = x + 5$

10-5 Study Guide and Intervention

The Pythagorean Theorem

The Pythagorean Theorem The side opposite the right angle in a right triangle is called the **hypotenuse**. This side is always the longest side of a right triangle. The other two sides are called the **legs** of the triangle. To find the length of any side of a right triangle, given the lengths of the other two sides, you can use the **Pythagorean Theorem**.

| **Pythagorean Theorem** | If a and b are the measures of the legs of a right triangle and c is the measure of the hypotenuse, then $c^2 = a^2 + b^2$. | 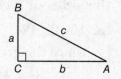 |

Example   **Find the missing length.**

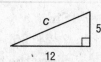

$$c^2 = a^2 + b^2 \qquad \text{Pythagorean Theorem}$$

$$c^2 = 5^2 + 12^2 \qquad a = 5 \text{ and } b = 12$$

$$c^2 = 169 \qquad \text{Simplify.}$$

$$c = \sqrt{169} \qquad \text{Take the square root of each side.}$$

$$c = 13 \qquad \text{Simplify.}$$

The length of the hypotenuse is 13.

Exercises

Find the length of each missing side. If necessary, round to the nearest hundredth.

1.

2.
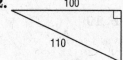

3.

4.
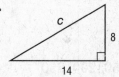

5.

6.
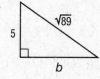

10-5 **Study Guide and Intervention** *(continued)*

The Pythagorean Theorem

Right Triangles If a and b are the measures of the shorter sides of a triangle, c is the measure of the longest side, and $c^2 = a^2 + b^2$, then the triangle is a right triangle.

Example Determine whether each set of measures can be sides of a right triangle.

a. 10, 12, 14

Since the greatest measure is 14, let $c = 14$, $a = 10$, and $b = 12$.

$c^2 = a^2 + b^2$ Pythagorean Theorem

$14^2 \stackrel{?}{=} 10^2 + 12^2$ $a = 10, b = 12, c = 14$

$196 \stackrel{?}{=} 100 + 144$ Multiply.

$196 \neq 244$ Add.

Since $c^2 \neq a^2 + b^2$, segments with these measures cannot form a right triangle.

b. 7, 24, 25

Since the greatest measure is 25, let $c = 25$, $a = 7$, and $b = 24$.

$c^2 = a^2 + b^2$ Pythagorean Theorem

$25^2 \stackrel{?}{=} 7^2 + 24^2$ $a = 7, b = 24, c = 25$

$625 \stackrel{?}{=} 49 + 576$ Multiply.

$625 = 625$ Add.

Since $c^2 = a^2 + b^2$, segments with these measures can form a right triangle.

Exercises

Determine whether each set of measures can be sides of a right triangle. Then determine whether they form a Pythagorean triple.

1. 14, 48, 50   **2.** 6, 8, 10   **3.** 8, 8, 10

4. 90, 120, 150   **5.** 15, 20, 25   **6.** 4, 8, $4\sqrt{5}$

7. 2, 2, $\sqrt{8}$ **8.** 4, 4, $\sqrt{20}$ **9.** 25, 30, 35

10. 24, 36, 48   **11.** 18, 80, 82   **12.** 150, 200, 250

13. 100, 200, 300   **14.** 500, 1200, 1300   **15.** 700, 1000, 1300

10-6 Study Guide and Intervention

Trigonometric Ratios

Trigonometric Ratios Trigonometry is the study of relationships of the angles and the sides of a right triangle. The three most common trigonometric ratios are the **sine**, **cosine**, and **tangent**.

sine of $\angle A = \dfrac{\text{leg opposite } \angle A}{\text{hypotenuse}}$	$\sin A = \dfrac{a}{c}$	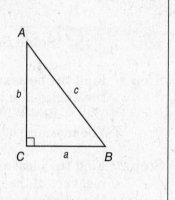
sine of $\angle B = \dfrac{\text{leg opposite } \angle B}{\text{hypotenuse}}$	$\sin B = \dfrac{b}{c}$	
cosine of $\angle A = \dfrac{\text{leg adjacent to } \angle A}{\text{hypotenuse}}$	$\cos A = \dfrac{b}{c}$	
cosine of $\angle B = \dfrac{\text{leg adjacent to } \angle B}{\text{hypotenuse}}$	$\cos B = \dfrac{a}{c}$	
tangent of $\angle A = \dfrac{\text{leg opposite } \angle A}{\text{leg adjacent to } \angle A}$	$\tan A = \dfrac{a}{b}$	
tangent of $\angle B = \dfrac{\text{leg opposite } \angle B}{\text{leg adjacent to } \angle B}$	$\tan B = \dfrac{b}{a}$	

Example   **Find the values of the three trigonometric ratios for angle A.**

Step 1 Use the Pythagorean Theorem to find BC.

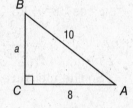

$a^2 + b^2 = c^2$ Pythagorean Theorem

$a^2 + 8^2 = 10^2$ $b = 8$ and $c = 10$

$a^2 + 64 = 100$ Simplify.

$a^2 = 36$ Subtract 64 from each side.

$a = 6$ Take the positive square root of each side.

Step 2 Use the side lengths to write the trigonometric ratios.

$$\sin A = \frac{\text{opp}}{\text{hyp}} = \frac{6}{10} = \frac{3}{5} \qquad \cos A = \frac{\text{adj}}{\text{hyp}} = \frac{8}{10} = \frac{4}{5} \qquad \tan A = \frac{\text{opp}}{\text{adj}} = \frac{6}{8} = \frac{3}{4}$$

Exercises

Find the values of the three trigonometric ratios for angle A.

1.

2.

3.

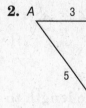

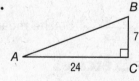

Use a calculator to find the value of each trigonometric ratio to the nearest ten-thousandth.

4. $\sin 40°$ **5.** $\cos 25°$ **6.** $\tan 85°$

10-6 Study Guide and Intervention (continued)

Trigonometric Ratios

Use Trigonometric Ratios When you find all of the unknown measures of the sides and angles of a right triangle, you are **solving the triangle**. You can find the missing measures of a right triangle if you know the measure of two sides of the triangle, or the measure of one side and the measure of one acute angle.

Example Solve the right triangle. Round each side length to the nearest tenth.

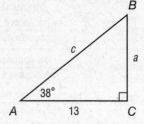

Step 1 Find the measure of ∠B. The sum of the measures of the angles in a triangle is 180.
$180° − (90° + 38°) = 52°$
The measure of ∠B is 52°.

Step 2 Find the measure of $\overline{AB}$. Because you are given the measure of the side adjacent to ∠A and are finding the measure of the hypotenuse, use the cosine ratio.

$\cos 38° = \dfrac{13}{c}$ Definition of cosine

$c \cos 38° = 13$ Multiply each side by c.

$c = \dfrac{13}{\cos 38°}$ Divide each side by cos 38°.

So the measure of $\overline{AB}$ is about 16.5.

Step 3 Find the measure of $\overline{BC}$. Because you are given the measure of the side adjacent to ∠A and are finding the measure of the side opposite ∠A, use the tangent ratio.

$\tan 38° = \dfrac{a}{13}$ Definition of tangent

$13 \tan 38° = a$ Multiply each side by 13.

$10.2 \approx a$ Use a calculator.

So the measure of $\overline{BC}$ is about 10.2.

Exercises

Solve each right triangle. Round each side length to the nearest tenth.

1.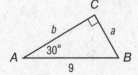

2.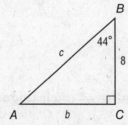

3.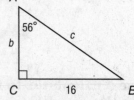

11-1 Study Guide and Intervention

Inverse Variation

Identify and Use Inverse Variations An **inverse variation** is an equation in the form of $y = \frac{k}{x}$ or $xy = k$. If two points (x_1, y_1) and (x_2, y_2) are solutions of an inverse variation, then $x_1 \cdot y_1 = k$ and $x_2 \cdot y_2 = k$.

Product Rule for Inverse Variation	$x_1 \cdot y_1 = x_2 \cdot y_2$

From the product rule, you can form the proportion $\dfrac{x_1}{x_2} = \dfrac{y_2}{y_1}$.

Example   If y varies inversely as x and $y = 12$ when $x = 4$, find x when $y = 18$.

Method 1 Use the product rule.

$$x_1 \cdot y_1 = x_2 \cdot y_2 \qquad \text{Product rule for inverse variation}$$
$$4 \cdot 12 = x_2 \cdot 18 \qquad x_1 = 4,\ y_1 = 12,\ y_2 = 18$$
$$\frac{48}{18} = x_2 \qquad \text{Divide each side by 18.}$$
$$\frac{8}{3} = x_2 \qquad \text{Simplify.}$$

Method 2 Use a proportion.

$$\frac{x_1}{x_2} = \frac{y_2}{y_1} \qquad \text{Proportion for inverse variation}$$
$$\frac{4}{x_2} = \frac{18}{12} \qquad x_1 = 4,\ y_1 = 12,\ y_2 = 18$$
$$48 = 18x_2 \qquad \text{Cross multiply.}$$
$$\frac{8}{3} = x_2 \qquad \text{Simplify.}$$

Both methods show that $x_2 = \frac{8}{3}$ when $y = 18$.

Exercises

Determine whether each table or equation represents an *inverse* or a *direct* variation. Explain.

1.

x	y
3	6
5	10
8	16
12	24

2. $y = 6x$

3. $xy = 15$

Assume that y varies inversely as x. Write an inverse variation equation that relates x and y. Then solve.

4. If $y = 10$ when $x = 5$, find y when $x = 2$.

5. If $y = 8$ when $x = -2$, find y when $x = 4$.

6. If $y = 100$ when $x = 120$, find x when $y = 20$.

7. If $y = -16$ when $x = 4$, find x when $y = 32$.

8. If $y = -7.5$ when $x = 25$, find y when $x = 5$.

9. DRIVING The Gerardi family can travel to Oshkosh, Wisconsin, from Chicago, Illinois, in 4 hours if they drive an average of 45 miles per hour. How long would it take them if they increased their average speed to 50 miles per hour?

10. GEOMETRY For a rectangle with given area, the width of the rectangle varies inversely as the length. If the width of the rectangle is 40 meters when the length is 5 meters, find the width of the rectangle when the length is 20 meters.

11-1 Study Guide and Intervention (continued)

Inverse Variation

Graph Inverse Variations Situations in which the values of y decrease as the values of x increase are examples of **inverse variation**. We say that y varies inversely as x, or y is inversely proportional to x.

Inverse Variation Equation	an equation of the form $xy = k$, where $k \neq 0$

Example 1 Suppose you drive 200 miles without stopping. The time it takes to travel a distance varies inversely as the rate at which you travel. Let x = speed in miles per hour and y = time in hours. Graph the variation.

The equation $xy = 200$ can be used to represent the situation. Use various speeds to make a table.

x	y
10	20
20	10
30	6.7
40	5
50	4
60	3.3

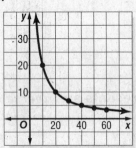

Example 2 Graph an inverse variation in which y varies inversely as x and $y = 3$ when $x = 12$.

Solve for k.

$$xy = k \qquad \text{Inverse variation equation}$$
$$12(3) = k \qquad x = 12 \text{ and } y = 3$$
$$36 = k \qquad \text{Simplify.}$$

Choose values for x and y, which have a product of 36.

x	y
−6	−6
−3	−12
−2	−18
2	18
3	12
6	6

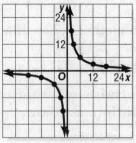

Exercises

Graph each variation if y varies inversely as x.

1. $y = 9$ when $x = -3$

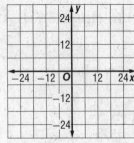

2. $y = 12$ when $x = 4$

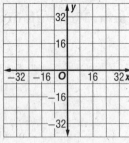

3. $y = -25$ when $x = 5$

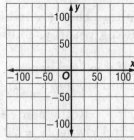

4. $y = 4$ when $x = 5$

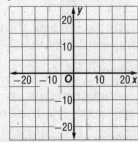

5. $y = -18$ when $x = -9$

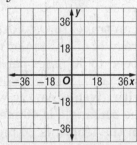

6. $y = 4.8$ when $x = 5.4$

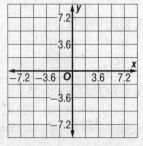

11-2 Study Guide and Intervention

Rational Functions

Identify Excluded Values The function $y = \dfrac{10}{x}$ is an example of a **rational function**. Because division by zero is undefined, any value of a variable that results in a denominator of zero must be excluded from the domain of that variable. These are called **excluded values** of the rational function.

Example   **State the excluded value for each function.**

a. $y = \dfrac{3}{x}$

    The denominator cannot equal zero.
    The excluded value is $x = 0$.

b. $y = \dfrac{4}{x - 5}$

    $x - 5 = 0$ Set the denominator equal to 0.

    $x = 5$ Add 5 to each side.

    The excluded value is $x = 5$.

Exercises

State the excluded value for each function.

1. $y = \dfrac{2}{x}$

2. $y = \dfrac{1}{x - 4}$

3. $y = \dfrac{x - 3}{x + 1}$

4. $y = \dfrac{4}{x - 2}$

5. $y = \dfrac{x}{2x - 4}$

6. $y = -\dfrac{5}{3x}$

7. $y = \dfrac{3x - 2}{x + 3}$

8. $y = \dfrac{x - 1}{5x + 10}$

9. $y = \dfrac{x + 1}{x}$

10. $y = \dfrac{x - 7}{2x + 8}$

11. $y = \dfrac{x - 5}{6x}$

12. $y = \dfrac{x - 2}{x + 11}$

13. $y = \dfrac{7}{3x + 21}$

14. $y = \dfrac{3x - 4}{x + 4}$

15. $y = \dfrac{x}{7x - 35}$

16. DINING Mya and her friends are eating at a restaurant. The total bill of $36 is split among x friends. The amount each person pays y is given by $y = \dfrac{36}{x}$, where x is the number of people. Graph the function.

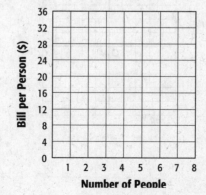

11-2 Study Guide and Intervention (continued)
Rational Functions

Identify and Use Asymptotes Because excluded vales are undefined, they affect the graph of the function. An **asymptote** is a line that the graph of a function approaches. A rational function in the form $y = \dfrac{a}{x - b} + c$ has a vertical asymptote at the x-value that makes the denominator equal zero, $x = b$. It has a horizontal asymptote at $y = c$.

> **Example** Identify the asymptotes of $y = \dfrac{1}{x - 1} + 2$. Then graph the function.

Step 1 Identify and graph the asymptotes using dashed lines.
vertical asymptote: $x = 1$
horizontal asymptote: $y = 2$

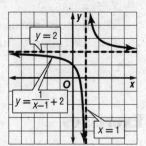

Step 2 Make a table of values and plot the points. Then connect them.

x	−1	0	2	3
y	1.5	1	3	2.5

Exercises

Identify the asymptotes of each function. Then graph the function.

1. $y = \dfrac{3}{x}$

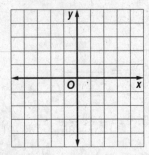

2. $y = \dfrac{-2}{x}$

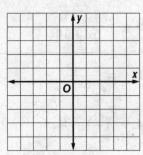

3. $y = \dfrac{4}{x} + 1$

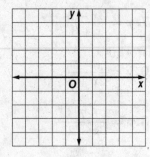

4. $y = \dfrac{2}{x} - 3$

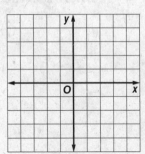

5. $y = \dfrac{2}{x + 1}$

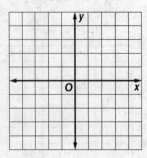

6. $y = \dfrac{-2}{x - 3}$

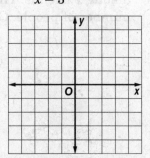

11-3 Study Guide and Intervention

Simplifying Rational Expressions

Identify Excluded Values

Rational Expression	an algebraic fraction with numerator and denominator that are polynomials	Example: $\dfrac{x^2 + 1}{y^2}$

Because a rational expression involves division, the denominator cannot equal zero. Any value of the denominator that results in division by zero is called an **excluded value** of the denominator.

Example 1 State the excluded value of $\dfrac{4m - 8}{m + 2}$.

Exclude the values for which $m + 2 = 0$.

$m + 2 = 0$ The denominator cannot equal 0.

$m + 2 - 2 = 0 - 2$ Subtract 2 from each side.

$m = -2$ Simplify.

Therefore, m cannot equal -2.

Example 2 State the excluded values of $\dfrac{x^2 + 1}{x^2 - 9}$.

Exclude the values for which $x^2 - 9 = 0$.

$x^2 - 9 = 0$ The denominator cannot equal 0.

$(x + 3)(x - 3) = 0$ Factor.

$x + 3 = 0$ or $x - 3 = 0$ Zero Product Property

$\quad\quad = -3 \quad\quad\quad\quad = 3$

Therefore, x cannot equal -3 or 3.

Exercises

State the excluded values for each rational expression.

1. $\dfrac{2b}{b^2 - 8}$

2. $\dfrac{12 - a}{32 + a}$

3. $\dfrac{x^2 - 2}{x^2 + 4}$

4. $\dfrac{m^2 - 4}{2m^2 - 8}$

5. $\dfrac{2n - 12}{n^2 - 4}$

6. $\dfrac{2x + 18}{x^2 - 16}$

7. $\dfrac{x^2 - 4}{x^2 + 4x + 4}$

8. $\dfrac{a - 1}{a^2 + 5a + 6}$

9. $\dfrac{k^2 - 2k + 1}{k^2 + 4k + 3}$

10. $\dfrac{m^2 - 1}{2m^2 - m - 1}$

11. $\dfrac{25 - n^2}{n^2 - 4n - 5}$

12. $\dfrac{2x^2 + 5x + 1}{x^2 - 10x + 16}$

13. $\dfrac{n^2 - 2n - 3}{n^2 + 4n - 5}$

14. $\dfrac{y^2 - y - 2}{3y^2 - 12}$

15. $\dfrac{k^2 + 2k - 3}{k^2 - 20k + 64}$

16. $\dfrac{x^2 + 4x + 4}{4x^2 + 11x - 3}$

11-3 Study Guide and Intervention (continued)

Simplifying Rational Expressions

Simplify Expressions Factoring polynomials is a useful tool for simplifying rational expressions. To simplify a rational expression, first factor the numerator and denominator. Then divide each by the greatest common factor.

Example 1 Simplify $\dfrac{54z^3}{24yz}$.

$$\dfrac{54z^3}{24yz} = \dfrac{(6z)(9z^2)}{(6z)(4y)}$$ The GCF of the numerator and the denominator is 6z.

$$= \dfrac{\cancel{(6z)}^1(9z^2)}{\cancel{(6z)}^1(4y)}$$ Divide the numerator and denominator by 6z.

$$= \dfrac{9z^2}{4y}$$ Simplify.

Example 2 Simplify $\dfrac{3x-9}{x^2-5x+6}$. **State the excluded values of x.**

$$\dfrac{3x-9}{x^2-5x+6} = \dfrac{3(x-3)}{(x-2)(x-3)}$$ Factor.

$$= \dfrac{3\cancel{(x-3)}^1}{(x-2)\cancel{(x-3)}_1}$$ Divide by the GCF, $x-3$.

$$= \dfrac{3}{x-2}$$ Simplify.

Exclude the values for which $x^2 - 5x + 6 = 0$.

$$x^2 - 5x + 6 = 0$$
$$(x-2)(x-3) = 0$$
$$x = 2 \quad \text{or} \quad x = 3$$

Therefore, $x \neq 2$ and $x \neq 3$.

Exercises

Simplify each expression. State the excluded values of the variables.

1. $\dfrac{12ab}{a^2b^2}$

2. $\dfrac{7n^3}{21n^8}$

3. $\dfrac{x+2}{x^2-4}$

4. $\dfrac{m^2-4}{m^2+6m+8}$

5. $\dfrac{2n-8}{n^2-16}$

6. $\dfrac{x^2+2x+1}{x^2-1}$

7. $\dfrac{x^2-4}{x^2+4x+4}$

8. $\dfrac{a^2+3a+2}{a^2+5a+6}$

9. $\dfrac{k^2-1}{k^2+4k+3}$

10. $\dfrac{m^2-2m+1}{2m^2-m-1}$

11. $\dfrac{n^2-25}{n^2-4n-5}$

12. $\dfrac{x^2+x-6}{2x^2-8}$

13. $\dfrac{n^2+7n+12}{n^2+2n-8}$

14. $\dfrac{y^2-y-2}{y^2-10y+16}$

11-4 Study Guide and Intervention

Multiplying and Dividing Rational Expressions

Multiply Rational Expressions To multiply rational expressions, you multiply the numerators and multiply the denominators. Then simplify.

Example 1 Find $\dfrac{2c^2f}{5ab^2} \cdot \dfrac{a^2b}{3cf}$.

$$\dfrac{2c^2f}{5ab^2} \cdot \dfrac{a^2b}{3cf} = \dfrac{2a^2bc^2f}{15ab^2cf} \qquad \text{Multiply.}$$

$$= \dfrac{{}^1(abcf)(2ac)}{{}^1(abcf)(15b)} \qquad \text{Simplify.}$$

$$= \dfrac{2ac}{15b} \qquad \text{Simplify.}$$

Example 2 Find $\dfrac{x^2 - 16}{2x + 8} \cdot \dfrac{x + 4}{x^2 + 8x + 16}$.

$$\dfrac{x^2 - 16}{2x + 8} \cdot \dfrac{x + 4}{x^2 + 8x + 16} = \dfrac{(x - 4)(x + 4)}{2(x + 4)} \cdot \dfrac{x + 4}{(x + 4)(x + 4)} \qquad \text{Factor.}$$

$$= \dfrac{(x - 4)(x + 4)^1}{2(x + 4)_1} \cdot \dfrac{x + 4^1}{(x + 4)(x + 4)_1} \qquad \text{Simplify.}$$

$$= \dfrac{x - 4}{2x + 8} \qquad \text{Multiply.}$$

Exercises

Find each product.

1. $\dfrac{6ab}{a^2b^2} \cdot \dfrac{a^2}{b^2}$

2. $\dfrac{mp^2}{3} \cdot \dfrac{4}{mp}$

3. $\dfrac{x + 2}{x - 4} \cdot \dfrac{x - 4}{x - 1}$

4. $\dfrac{m - 5}{8} \cdot \dfrac{16}{m - 5}$

5. $\dfrac{2n - 8}{n + 2} \cdot \dfrac{2n + 4}{n - 4}$

6. $\dfrac{x^2 - 64}{2x + 16} \cdot \dfrac{x + 8}{x^2 + 16x + 64}$

7. $\dfrac{8x + 8}{x^2 - 2x + 1} \cdot \dfrac{x - 1}{2x + 2}$

8. $\dfrac{a^2 - 25}{a + 2} \cdot \dfrac{a^2 - 4}{a - 5}$

9. $\dfrac{x^2 + 6x + 8}{2x^2 + 9x + 4} \cdot \dfrac{2x^2 - x - 1}{x^2 - 3x + 2}$

10. $\dfrac{m^2 - 1}{2m^2 - m - 1} \cdot \dfrac{2m + 1}{m^2 - 2m + 1}$

11. $\dfrac{n^2 - 1}{n^2 - 7n + 10} \cdot \dfrac{n^2 - 25}{n^2 + 6n + 5}$

12. $\dfrac{3p - 3r}{10pr} \cdot \dfrac{20p^2r^2}{p^2 - r^2}$

13. $\dfrac{a^2 + 7a + 12}{a^2 + 2a - 8} \cdot \dfrac{a^2 + 3a - 10}{a^2 + 2a - 8}$

14. $\dfrac{v^2 - 4v - 21}{3v^2 + 6v} \cdot \dfrac{v^2 + 8v}{v^2 + 11v + 24}$

11-4 Study Guide and Intervention (continued)

Multiplying and Dividing Rational Expressions

Divide Rational Expressions To divide rational expressions, multiply by the reciprocal of the divisor. Then simplify.

Example 1 Find $\dfrac{12c^2f}{5a^2b^2} \div \dfrac{c^2f^2}{10ab}$.

$$\dfrac{12c^2f}{5a^2b^2} \div \dfrac{c^2f^2}{10ab} = \dfrac{12c^2f}{5a^2b^2} \times \dfrac{10ab}{c^2f^2}$$

$$= \dfrac{12^1 c^2 f^1}{{}_1 5 a^2 b^2{}_b} \times \dfrac{{}^2 10^1 ab^1}{{}_1 c^2 f^2{}_f}$$

$$= \dfrac{24}{abf}$$

Example 2 Find $\dfrac{x^2 + 6x - 27}{x^2 + 11x + 18} \div \dfrac{x - 3}{x^2 + x - 2}$.

$$\dfrac{x^2 + 6x - 27}{x^2 + 11x + 18} \div \dfrac{x - 3}{x^2 + x - 2} = \dfrac{x^2 + 6x - 27}{x^2 + 11x + 18} \times \dfrac{x^2 + x - 2}{x - 3}$$

$$= \dfrac{(x + 9)(x - 3)}{(x + 9)(x + 2)} \times \dfrac{(x + 2)(x - 1)}{x - 3}$$

$$= \dfrac{{}^1(x + 9)(x - 3)^1}{{}_1(x + 9)(x + 2)_1} \times \dfrac{{}^1(x + 2)(x - 1)}{x - 3_1}$$

$$= x - 1$$

Exercises

Find each quotient.

1. $\dfrac{12ab}{a^2b^2} \div \dfrac{b}{a}$

2. $\dfrac{n}{4} \div \dfrac{n}{p}$

3. $\dfrac{3xy^2}{8} \div 6xy$

4. $\dfrac{m - 5}{8} \div \dfrac{m - 5}{16}$

5. $\dfrac{2n - 4}{2n} \div \dfrac{n^2 - 4}{n}$

6. $\dfrac{y^2 - 36}{y^2 - 49} \div \dfrac{y + 6}{y + 7}$

7. $\dfrac{x^2 - 5x + 6}{5} \div \dfrac{x - 3}{15}$

8. $\dfrac{a^2b^3c}{3r^2t} \div \dfrac{6a^2bc}{8rt^2u}$

9. $\dfrac{x^2 + 6x + 8}{x^2 + 4x + 4} \div \dfrac{x + 4}{x + 2}$

10. $\dfrac{m^2 - 49}{m} \div \dfrac{m^2 - 13m + 42}{3m^2}$

11. $\dfrac{n^2 - 5n + 6}{n^2 + 3n} \div \dfrac{3 - n}{4n + 12}$

12. $\dfrac{p^2 - 2pr + r^2}{p + r} \div \dfrac{p^2 - r^2}{p + r}$

13. $\dfrac{a^2 + 7a + 12}{a^2 + 3a - 10} \div \dfrac{a^2 - 9}{a^2 - 25}$

14. $\dfrac{a^2 - 9}{2a^2 + 13a - 7} \div \dfrac{a + 3}{4a^2 - 1}$

11-5 Study Guide and Intervention

Dividing Polynomials

Divide Polynomials by Monomials
To divide a polynomial by a monomial, divide each term of the polynomial by the monomial.

Example 1 Find $(4r^2 - 12r) \div (2r)$.

$$(4r^2 - 12r) \div 2r = \frac{4r^2 - 12r}{2r}$$

$$= \frac{4r^2}{2r} - \frac{12r}{2r} \quad \text{Divide each term.}$$

$$= \frac{^{2r}\cancel{4r^2}}{_1\cancel{2r}} - \frac{\cancel{12r}^{\,6}}{\cancel{2r}_{\,1}} \quad \text{Simplify.}$$

$$= 2r - 6 \quad \text{Simplify.}$$

Example 2 Find $(3x^2 - 8x + 4) \div (4x)$.

$$(3x^2 - 8x + 4) \div 4x = \frac{3x^2 - 8x + 4}{4x}$$

$$= \frac{3x^2}{4x} - \frac{8x}{4x} + \frac{4}{4x}$$

$$= \frac{^{3x}\cancel{3x^2}}{_4\cancel{4x}} - \frac{^2\cancel{8x}}{^1\cancel{4x}} + \frac{4}{4x}$$

$$= \frac{3x}{4} - 2 + \frac{1}{x}$$

Exercises

Find each quotient.

1. $(x^3 + 2x^2 - x) \div x$

2. $(2x^3 + 12x^2 - 8x) \div (2x)$

3. $(x^2 + 3x - 4) \div x$

4. $(4m^2 + 6m - 8) \div (2m^2)$

5. $(3x^3 + 15x^2 - 21x) \div (3x)$

6. $(8m^2p^2 + 4mp - 8p) \div p$

7. $(8y^4 + 16y^2 - 4) \div (4y^2)$

8. $(16x^4y^2 + 24xy + 5) \div (xy)$

9. $\dfrac{15x^2 - 25x + 30}{5}$

10. $\dfrac{10a^2b + 12ab - 8b}{2a}$

11. $\dfrac{6x^3 + 9x^2 + 9}{3x}$

12. $\dfrac{m^2 - 12m + 42}{3m^2}$

13. $\dfrac{m^2p^2 - 5mp + 6}{m^2p^2}$

14. $\dfrac{p^2 - 4pr + 6r^2}{pr}$

15. $\dfrac{6a^2b^2 - 8ab + 12}{2a^2}$

16. $\dfrac{2x^2y^3 - 4x^2y^2 - 8xy}{2xy}$

17. $\dfrac{9x^2y^2z - 2xyz + 12x}{xy}$

18. $\dfrac{2a^3b^3 + 8a^2b^2 - 10ab + 12}{2a^2b^2}$

11-5 Study Guide and Intervention (continued)

Dividing Polynomials

Divide Polynomials by Binomials To divide a polynomial by a binomial, factor the dividend if possible and divide both dividend and divisor by the GCF. If the polynomial cannot be factored, use long division.

> **Example** Find $(x^2 + 7x + 10) \div (x + 3)$.

Step 1 Divide the first term of the dividend, x^2 by the first term of the divisor, x.

$$
\begin{array}{r}
x \\
x+3\overline{)x^2 + 7x + 10} \\
\underline{(-)\,x^2 + 3x} \\
4x
\end{array}
$$

Multiply x and $x + 3$.

Subtract.

Step 2 Bring down the next term, 10. Divide the first term of $4x + 10$ by x.

$$
\begin{array}{r}
x + 4 \\
x+3\overline{)x^2 + 7x + 10} \\
\underline{x^2 + 3x} \\
4x + 10 \\
\underline{(-)\,4x + 12} \\
-2
\end{array}
$$

Multiply 4 and $x + 3$.

Subtract.

The quotient is $x + 4$ with remainder -2. The quotient can be written as $x + 4 + \dfrac{-2}{x + 3}$.

Exercises

Find each quotient.

1. $(b^2 - 5b + 6) \div (b - 2)$

2. $(x^2 - x - 6) \div (x - 3)$

3. $(x^2 + 3x - 4) \div (x - 1)$

4. $(m^2 + 2m - 8) \div (m + 4)$

5. $(x^2 + 5x + 6) \div (x + 2)$

6. $(m^2 + 4m + 4) \div (m + 2)$

7. $(2y^2 + 5y + 2) \div (y + 2)$

8. $(8y^2 - 15y - 2) \div (y - 2)$

9. $\dfrac{8x^2 - 6x - 9}{4x + 3}$

10. $\dfrac{m^2 - 5m - 6}{m - 6}$

11. $\dfrac{x^3 + 1}{x - 2}$

12. $\dfrac{6m^3 + 11m^2 + 4m + 35}{2m + 5}$

13. $\dfrac{6a^2 + 7a + 5}{2a + 5}$

14. $\dfrac{8p^3 + 27}{2p + 3}$

11-6 Study Guide and Intervention

Adding and Subtracting Rational Expressions

Add and Subtract Rational Expressions with Like Denominators To add rational expressions with like denominators, add the numerators and then write the sum over the common denominator. To subtract fractions with like denominators, subtract the numerators. If possible, simplify the resulting rational expression.

Example 1 Find $\dfrac{5n}{15} + \dfrac{7n}{15}$.

$\dfrac{5n}{15} + \dfrac{7n}{15} = \dfrac{5n + 7n}{15}$ Add the numerators.

$= \dfrac{12n}{15}$ Simplify.

$= \dfrac{\overset{4n}{\cancel{12n}}}{\underset{5}{\cancel{15}}}$ Divide by 3.

$= \dfrac{4n}{5}$ Simplify.

Example 2 Find $\dfrac{3x + 2}{x - 2} - \dfrac{4x}{x - 2}$.

$\dfrac{3x + 2}{x - 2} - \dfrac{4x}{x - 2} = \dfrac{3x + 2 - 4x}{x - 2}$ The common

denominator is $x - 2$.

$= \dfrac{2 - x}{x - 2}$ Subtract.

$= \dfrac{-1(x - 2)}{x - 2}$ $2 - x = -1(x - 2)$

$= \dfrac{-1\overset{1}{\cancel{(x - 2)}}}{\underset{1}{\cancel{x - 2}}}$

$= \dfrac{-1}{1}$ Simplify.

$= -1$

Exercises

Find each sum or difference.

1. $\dfrac{3}{a} + \dfrac{4}{a}$

2. $\dfrac{x^2}{8} + \dfrac{x}{8}$

3. $\dfrac{5x}{9} - \dfrac{x}{9}$

4. $\dfrac{11x}{15y} - \dfrac{x}{15y}$

5. $\dfrac{2a - 4}{a - 4} + \dfrac{-a}{a - 4}$

6. $\dfrac{m + 1}{2m - 1} + \dfrac{3m - 3}{2m - 1}$

7. $\dfrac{y + 7}{y + 6} - \dfrac{1}{y + 6}$

8. $\dfrac{3y + 5}{5} - \dfrac{2y}{5}$

9. $\dfrac{x + 1}{x - 2} + \dfrac{x - 5}{x - 2}$

10. $\dfrac{5a}{3b^2} + \dfrac{10a}{3b^2}$

11. $\dfrac{x^2 + x}{x} - \dfrac{x^2 + 5x}{x}$

12. $\dfrac{5a + 2}{a^2} - \dfrac{4a + 2}{a^2}$

13. $\dfrac{3x + 2}{x + 2} + \dfrac{x + 6}{x + 2}$

14. $\dfrac{a - 4}{a + 1} + \dfrac{a + 6}{a + 1}$

11-6 Study Guide and Intervention (continued)

Adding and Subtracting Rational Expressions

Add and Subtract Rational Expressions with Unlike Denominators Adding or subtracting rational expressions with unlike denominators is similar to adding and subtracting fractions with unlike denominators.

Adding and Subtracting Rational Expressions	Step 1 Find the LCD of the expressions.
	Step 2 Change each expression into an equivalent expression with the LCD as the denominator.
	Step 3 Add or subtract just as with expressions with like denominators.
	Step 4 Simplify if necessary.

Example 1 Find $\dfrac{n+3}{n} + \dfrac{8n-4}{4n}$.

Factor each denominator.

$n = n$

$4n = 4 \cdot n$

$\text{LCD} = 4n$

Since the denominator of $\dfrac{8n-4}{4n}$ is already $4n$, only $\dfrac{n+3}{n}$ needs to be renamed.

$\dfrac{n+3}{n} + \dfrac{8n-4}{4n} = \dfrac{4(n+3)}{4n} + \dfrac{8n-4}{4n}$

$= \dfrac{4n+12}{4n} + \dfrac{8n-4}{4n}$

$= \dfrac{12n+8}{4n}$

$= \dfrac{3n+2}{n}$

Example 2 Find $\dfrac{3x}{x^2-4x} - \dfrac{1}{x-4}$.

$\dfrac{3x}{x^2-4x} - \dfrac{1}{x-4} = \dfrac{3x}{x(x-4)} - \dfrac{1}{x-4}$ Factor the denominator.

$= \dfrac{3x}{x(x-4)} - \dfrac{1}{x-4} \cdot \dfrac{x}{x}$ The LCD is $x(x-4)$.

$= \dfrac{3x}{x(x-4)} - \dfrac{x}{x(x-4)}$ $1 \cdot x = x$

$= \dfrac{2x}{x(x-4)}$ Subtract numerators.

$= \dfrac{2}{x-4}$ Simplify.

Exercises

Find each sum or difference.

1. $\dfrac{1}{a} + \dfrac{7}{3a}$

2. $\dfrac{1}{6x} + \dfrac{3}{8}$

3. $\dfrac{5}{9x} - \dfrac{1}{x^2}$

4. $\dfrac{6}{x^2} - \dfrac{3}{x^3}$

5. $\dfrac{8}{4a^2} + \dfrac{6}{3a}$

6. $\dfrac{4}{h+1} + \dfrac{2}{h+2}$

7. $\dfrac{y}{y-3} - \dfrac{3}{y+3}$

8. $\dfrac{y}{y-7} - \dfrac{y+3}{y^2-4y-21}$

9. $\dfrac{a}{a+4} + \dfrac{4}{a-4}$

10. $\dfrac{6}{3(m+1)} + \dfrac{2}{3(m-1)}$

11. $\dfrac{4}{x-2y} - \dfrac{2}{x+2y}$

12. $\dfrac{a-6b}{2a^2-5ab+2b^2} - \dfrac{7}{a-2b}$

13. $\dfrac{y+2}{y^2+5y+6} + \dfrac{2-y}{y^2+y-6}$

14. $\dfrac{q}{q^2-16} + \dfrac{q+1}{q^2+5q+4}$

11-7 Study Guide and Intervention

Mixed Expressions and Complex Fractions

Simplify Mixed Expressions Algebraic expressions such as $a + \dfrac{b}{c}$ and $5 + \dfrac{x+y}{x+3}$ are called **mixed expressions**. Changing mixed expressions to rational expressions is similar to changing mixed numbers to improper fractions.

Example 1 Simplify $5 + \dfrac{2}{n}$.

$5 + \dfrac{2}{n} = \dfrac{5 \cdot n}{n} + \dfrac{2}{n}$ LCD is n.

$\phantom{5 + \dfrac{2}{n}} = \dfrac{5n + 2}{n}$ Add the numerators.

Therefore, $5 + \dfrac{2}{n} = \dfrac{5n + 2}{n}$.

Example 2 Simplify $2 + \dfrac{3}{n+3}$.

$2 + \dfrac{3}{n+3} = \dfrac{2(n+3)}{n+3} + \dfrac{3}{n+3}$

$\phantom{2 + \dfrac{3}{n+3}} = \dfrac{2n+6}{n+3} + \dfrac{3}{n+3}$

$\phantom{2 + \dfrac{3}{n+3}} = \dfrac{2n+6+3}{n+3}$

$\phantom{2 + \dfrac{3}{n+3}} = \dfrac{2n+9}{n+3}$

Therefore, $2 + \dfrac{3}{n+3} = \dfrac{2n+9}{n+3}$.

Exercises

Write each mixed expression as a rational expression.

1. $4 + \dfrac{6}{a}$

2. $\dfrac{1}{9x} - 3$

3. $3x - \dfrac{1}{x^2}$

4. $\dfrac{4}{x^2} - 2$

5. $10 + \dfrac{60}{x+5}$

6. $\dfrac{h}{h+4} + 2$

7. $\dfrac{y}{y-2} + y^2$

8. $4 - \dfrac{4}{2x+1}$

9. $1 + \dfrac{1}{x}$

10. $\dfrac{4}{m-2} - 2m$

11. $x^2 + \dfrac{x+2}{x-3}$

12. $a - 3 + \dfrac{a-2}{a+3}$

13. $4m + \dfrac{3p}{2t}$

14. $2q^2 + \dfrac{q}{p+q}$

15. $\dfrac{2}{y^2-1} - 4y^2$

16. $t^2 + \dfrac{p+t}{p-t}$

11-7 Study Guide and Intervention (continued)

Mixed Expressions and Complex Fractions

Simplify Complex Fractions If a fraction has one or more fractions in the numerator or denominator, it is called a **complex fraction**.

Simplifying a Complex Fraction	Any complex fraction $\dfrac{\frac{a}{b}}{\frac{c}{d}}$ where $b \neq 0$, $c \neq 0$, and $d \neq 0$, can be expressed as $\dfrac{ad}{bc}$.

Example Simplify $\dfrac{2 + \frac{4}{a}}{\frac{a+2}{3}}$.

$$\frac{2 + \frac{4}{a}}{\frac{a+2}{3}} = \frac{\frac{2a}{a} + \frac{4}{a}}{\frac{a+2}{3}}$$ Find the LCD for the numerator and rewrite as like fractions.

$$= \frac{\frac{2a+4}{a}}{\frac{a+2}{3}}$$ Simplify the numerator.

$$= \frac{2a+4}{a} \cdot \frac{3}{a+2}$$ Rewrite as the product of the numerator and the reciprocal of the denominator.

$$= \frac{2(a+2)}{a} \cdot \frac{3}{a+2}$$ Factor.

$$= \frac{6}{a}$$ Divide and simplify.

Exercises

Simplify each expression.

1. $\dfrac{2\frac{2}{5}}{3\frac{3}{4}}$

2. $\dfrac{\frac{3}{x}}{\frac{4}{y}}$

3. $\dfrac{\frac{x}{y^3}}{\frac{x^3}{y^2}}$

4. $\dfrac{1 - \frac{1}{x}}{1 + \frac{1}{x}}$

5. $\dfrac{1 - \frac{1}{x}}{1 - \frac{1}{x^2}}$

6. $\dfrac{\frac{1}{x-3}}{\frac{2}{x^2-9}}$

7. $\dfrac{\frac{x^2-25}{y}}{x^3-5x^2}$

8. $\dfrac{x - \frac{12}{x-1}}{x - \frac{8}{x-2}}$

9. $\dfrac{\frac{3}{y+2} - \frac{2}{y-2}}{\frac{1}{y+2} - \frac{2}{y-2}}$

11-8 Study Guide and Intervention

Rational Equations

Solve Rational Equations **Rational equations** are equations that contain rational expressions. To solve equations containing rational expressions, multiply each side of the equation by the least common denominator.

Rational equations can be used to solve **work problems** and **rate problems**.

Example 1 Solve $\dfrac{x-3}{3} + \dfrac{x}{2} = 4$.

$$\dfrac{x-3}{3} + \dfrac{x}{2} = 4$$

$6\left(\dfrac{x-3}{3} + \dfrac{x}{2}\right) = 6(4)$ The LCD is 6.

$2(x-3) + 3x = 24$ Distributive Property

$2x - 6 + 3x = 24$ Distributive Property

$5x = 30$ Simplify.

$x = 6$ Divide each side by 5.

The solution is 6.

Example 2 Solve $\dfrac{15}{x^2-1} = \dfrac{5}{2(x-1)}$. State any extraneous solutions.

$\dfrac{15}{x^2-1} = \dfrac{5}{2(x-1)}$ Original equation

$30(x-1) = 5(x^2-1)$ Cross multiply.

$30x - 30 = 5x^2 - 5$ Distributive Property

$0 = 5x^2 - 30x + 30 - 5$ Add $-30x + 30$ to each side.

$0 = 5x^2 - 30x + 25$ Simplify.

$0 = 5(x^2 - 6x + 5)$ Factor.

$0 = 5(x-1)(x-5)$ Factor.

$x = 1$ or $x = 5$ Zero Product Property

The number 1 is an extraneous solution, since 1 is an excluded value for x. So, 5 is the solution of the equation.

Exercises

Solve each equation. State any extraneous solutions.

1. $\dfrac{x-5}{5} + \dfrac{x}{4} = 8$

2. $\dfrac{3}{x} = \dfrac{6}{x+1}$

3. $\dfrac{x-1}{5} = \dfrac{2x-2}{15}$

4. $\dfrac{8}{n-1} = \dfrac{10}{n+1}$

5. $t - \dfrac{4}{t+3} = t + 3$

6. $\dfrac{m+4}{m} + \dfrac{m}{3} = \dfrac{m}{3}$

7. $\dfrac{q+4}{q-1} + \dfrac{q}{q+1} = 2$

8. $\dfrac{5-2x}{2} - \dfrac{4x+3}{6} = \dfrac{7x+2}{6}$

9. $\dfrac{m+1}{m-1} - \dfrac{m}{1-m} = 1$

10. $\dfrac{x^2-9}{x-3} + x^2 = 9$

11. $\dfrac{2}{x^2-36} - \dfrac{1}{x-6} = 0$

12. $\dfrac{4z}{z^2+4z+3} = \dfrac{6}{z+3} + \dfrac{4}{z+1}$

13. $\dfrac{4}{4-p} - \dfrac{p^2}{p-4} = 4$

14. $\dfrac{x^2-16}{x-4} + x^2 = 16$

11-8 Study Guide and Intervention (continued)

Rational Equations

Use Rational Equations to Solve Problems Rational equation can be used to solve **work problems** and **rate problems**.

Example **WORK PROBLEM** Marla can paint Percy's kitchen in 3 hours. Percy can paint it in 2 hours. Working together, how long will it take Marla and Percy to paint the kitchen?

In t hours, Marla completes $t \cdot \dfrac{1}{3}$ of the job and Percy completes $t \cdot \dfrac{1}{2}$ of the job. So an equation for completing the whole job is $\dfrac{t}{3} + \dfrac{t}{2} = 1$.

$$\frac{t}{3} + \frac{t}{2} = 1$$

$2t + 3t = 6$ Multiply each term by 6.

$5t = 6$ Add like terms.

$t = \dfrac{6}{5}$ Solve.

So it will take Marla and Percy $1\dfrac{1}{5}$ hours to paint the room if they work together.

Exercises

1. **GREETING CARDS** It takes Kenesha 45 minutes to prepare 20 greeting cards. It takes Paula 30 minutes to prepare the same number of cards. Working together at this rate, how long will it take them to prepare the cards?

2. **BOATING** A motorboat went upstream at 15 miles per hour and returned downstream at 20 miles per hour. How far did the boat travel one way if the round trip took 3.5 hours?

3. **FLOORING** Maya and Reginald are installing hardwood flooring. Maya can install flooring in a room in 4 hours. Reginald can install flooring in a room in 3 hours. How long would it take them if they worked together?

4. **BICYCLING** Stefan is bicyling on a bike trail at an average of 10 miles per hour. Erik starts bicycling on the same trail 30 minutes later. If Erik averages 16 miles per hour, how long will it take him to pass Stefan?

12-1 Study Guide and Intervention

Samples and Studies

Sampling A **population** consists of all of the members of a group of interest. Since it may be impractical to examine every member of a population, a **sample** or subset is sometimes selected to represent the population.

Sample	Definition
Simple random	Each member of the population has an equal chance of being selected.
Systematic	Members are selected according to a specified interval from a random starting point.
Self-selected	Members volunteer to be included.
Convenience	Members that are readily available are selected.
Stratified	The population is first divided into similar, nonoverlapping groups. Members are then randomly selected from each group.

Example **SCHOOL** The principal of a high school wanted to know if students in the school liked the attendance policy. He decided to survey the students in the third-hour study hall about whether they like the attendance policy. Fewer than one fourth of the students in the school have a study hall.

a. Identify the sample, and suggest a population from which it was selected.

The sample includes only those students in the third-hour study hall. The population is the entire student body.

b. Classify the sample as *simple, systematic, self-selected, convenience,* or *stratified*. Explain your reasoning.

This sample is a convenience sample because it is convenient to sample students in a study hall during a certain period of the day.

Exercises

Identify each sample, and suggest a population from which it was selected. Then classify the sample as *simple, systematic, self-selected, convenience,* or *stratified*. Explain your reasoning.

1. SHOPPING Every tenth person leaving a grocery store was asked if they would participate in a community survey.

2. GARDENING A gardener divided a lot into 25-square-foot sections. He then took two soil samples from each and tested the samples for mineral content.

3. SCHOOL The counselors of a high school sent out a survey to senior students with questions about their plans for college. 40% of the seniors sent responses back.

12-1 Study Guide and Intervention (continued)

Samples and Studies

Studies Information is collected from samples using one of the following study types.

Type	Definition
Survey	Data are collected from responses given by a sample regarding their characteristics, behaviors, or opinions.
Observational Study	Members of a sample are measured or observed without being affected by the study.
Experiment	The sample is divided into two groups: • an *experimental group* that undergoes a change, and • a *control group* that does not undergo the change. The effect on the experimental group is then compared to the control group.

Example Determine whether each situation describes a *survey*, an *observational study*, or an *experiment*. Explain your reasoning.

a. **SHOPPING** The manager of a department store wants to analyze customers' overall shopping experience. Checkout clerks distribute a questionnaire to all shoppers that make a purchase.

This is a survey. The data are gathered from responses given by members of the sample.

b. **MARKETING** A marketing group places new displays for a product in selected stores and then observes shoppers' reactions as they pass by a display.

This is an observational study. The data are gathered from observing the shoppers.

Exercises

Determine whether each situation describes a *survey*, an *observational study*, or an *experiment*. Explain your reasoning.

1. **UNIFORMS** An athletic trainer is testing a new fabric. He has half of the football team wear their normal uniforms and the other half wears uniforms made from the new fabric. He observes and analyzes the mobility of the players.

2. **FOOD** A researcher from a pet food manufacturer places multiple bowls of its several flavors of dog food in a room with 20 dogs. The researchers records which foods the dogs eat.

3. **EVALUATION** At the end of the school year, each teacher distributes 10 standard questions to each of their students that ask the students to rate their teachers' guidelines and classroom practices on a 5-point scale.

12-2 Study Guide and Intervention

Statistics and Parameters

Statistics and Parameters A **statistic** is a measure that describes a characteristic of a sample. A **parameter** is a measure that describes a characteristic of a population. Statistics can change from sample to sample while parameters do not.

> **Example** Identify the sample and the population for each situation. Then describe the sample statistic and the population parameter.

a. At a local supermarket, a random sample of 50 shoppers is selected. The median amount spent at the supermarket is calculated for the sample.

Sample: the group of 50 shoppers
Population: all shoppers at the supermarket
Sample statistic: median amount of time spent by shoppers in the sample
Population parameter: median amount by all shoppers spent at the supermarket

b. Every 20 minutes at a furniture factory, a finished sofa is pulled from the assembly line and checked for defects. The percentage of sofas in the sample which are defective is then calculated.

Sample: the sofas checked for defects
Population: all sofas manufactured
Sample statistic: percentage of defective sofas in the sample
Population parameter: percentage of all sofas which are defective

Exercises

Identify the sample and the population for each situation. Then describe the sample statistic and the population parameter.

1. WEATHER A meteorologist places ten weather stations in a county to measure rainfall. The mean annual rainfall is calculated for the sample.

2. BOTANY A scientist randomly selects 20 trees in a forest. The mean height of the 20 trees is then calculated.

3. POLITICS A political reporter randomly selects 25 congressional districts across the country. The mean number of votes cast in the 25 congressional districts is calculated.

12-2 Study Guide and Intervention (continued)

Statistics and Parameters

Statistical Analysis The **mean absolute deviation** is the average of the absolute values of the differences between the mean and each value in the data set. It is used to predict errors and judge how well the mean represents the data. The **standard deviation** is the calculated value that shows how data deviate from the mean of the set of data. The **variance** of data is the square of the standard deviation.

Example **EMPLOYMENT** Employees at a law firm keep track of how many hours they work each week: {44, 48, 44, 40, 59}.

a. Find and interpret the mean absolute deviation.

Step 1 Find the mean. For this set of data, the mean is 47.

Step 2 Find the sum of the absolute values of the differences between each value in the set of data and the mean.

$$|44 - 47| + |48 - 47| + |44 - 47| + |40 - 47| + |59 - 47| =$$
$$3 + 1 + 3 + 7 + 12 = 26$$

Step 3 Divide the sum by the number of values in the set of data:
$26 \div 5 = 5.2$

The mean absolute deviation of 5.2 indicates that the data, on average, are 5.2 hours away from the mean of 47 hours.

b. Find and interpret the standard deviation.

Step 1 To find the variance, square the difference between each number and the mean. Then divide by the number of values.

$$\sigma^2 = \frac{(44 - 47)^2 + (48 - 47)^2 + (44 - 47)^2 + (40 - 47)^2 + (59 - 47)^2}{5}$$

$$= \frac{(-3)^2 + (1)^2 + (-3)^2 + (-7)^2 + (12)^2}{5} = \frac{9 + 1 + 9 + 49 + 144}{5} = \frac{212}{5} \text{ or } 42.4$$

Step 2 The standard deviation is the square root of the variance.

$$\sqrt{\sigma^2} = \sqrt{42.4}$$

$$\sigma \approx 6.51$$

The standard deviation is approximately 6.51, which is small compared to the mean of 47. This suggests that most of the employees work relatively close to 47 hours per week.

Exercises

Find and interpret the mean absolute deviation and standard deviation of each set of data.

1. {2, 4, 6, 5, 5, 3, 5, 4}

2. {13, 10, 36, 48, 52}

12-3 Study Guide and Intervention

Distributions of Data

Describing Distributions A **distribution** of data shows the frequency of each possible data value. The shape of a distribution can be determined by looking at its histogram.

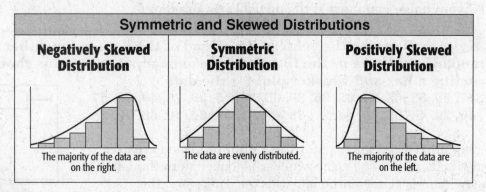

Symmetric and Skewed Distributions		
Negatively Skewed Distribution	**Symmetric Distribution**	**Positively Skewed Distribution**
The majority of the data are on the right.	The data are evenly distributed.	The majority of the data are on the left.

Example Use a graphing calculator to construct a histogram for the data, and use it to describe the shape of the distribution.

13, 38, 18, 20, 40, 22, 44, 24, 43, 39, 26, 28, 30, 43, 44, 33, 35, 32
36, 43, 16, 37, 38, 40, 25, 40, 31, 32, 25, 43, 23, 44, 39, 34, 46, 45

Use a graphing calculator to enter the data into L1 and create a histogram. Adjust the window to the dimensions shown. The graph is high on the right. Therefore, the distribution is negatively skewed.

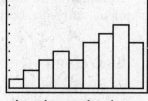

[12, 48] scl: 4 by [0, 10] scl: 1

Exercises

Use a graphing calculator to construct a histogram for the data, and use it to describe the shape of the distribution.

1. 52, 75, 77, 60, 79, 63,
55, 57, 64, 82, 65, 66,
70, 89, 88, 67, 71, 72

72, 74, 59, 75, 78, 80,
83, 66, 83, 62, 76, 68,
73, 71, 84, 54, 85, 92

2. 77, 61, 72, 65, 69, 62,
63, 75, 64, 89, 64, 86,
64, 74, 65, 71, 66, 67

69, 70, 80, 66, 71, 68,
74, 65, 75, 79, 68, 79,
63, 82, 62, 76, 84, 63

3. 46, 41, 45, 48, 49, 31,
47, 36, 48, 38, 47, 39,
49, 40, 48, 46, 34, 42

44, 47, 43, 37, 47, 45,
48, 42, 49, 44, 41, 50,
33, 43, 46, 37, 40, 43

12-3 Study Guide and Intervention (continued)

Distributions of Data

Analyzing Distributions When describing a distribution, use

* the mean and standard deviation if the distribution is symmetric, or
* the five-number summary if the distribution is skewed.

Example Describe the center and spread of the data using either the mean and standard deviation or the five-number summary. Justify your choice by constructing a box-and-whisker plot for the data.

53, 49, 47, 50, 53, 58, 34, 62, 35, 37, 65, 38, 35, 39, 73, 40, 70, 57
28, 41, 40, 26, 43, 33, 30, 44, 27, 39, 47, 31, 52, 60, 36, 68, 70, 28

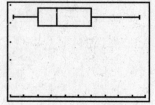

[25, 75] scl: 5 by [0, 5] scl: 1

Use a graphing calculator to enter the data into L1 and create a box-and-whisker plot. The right whisker is longer than the left and the median is closer to the left whisker. Therefore, the distribution is positively skewed.

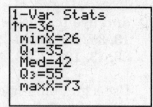

The distribution is positively skewed, so use the five-number summary. The range is 73 − 26 or 47. The median is 42, and half of the data are between 35 and 55.

Exercises

Describe the center and spread of the data using either the mean and standard deviation or the five-number summary. Justify your choice by constructing a box-and-whisker plot for the data.

1. 4, 25, 24, 2, 10, 11, 26, 28, 13, 25, 26, 14, 16, 17, 27, 29, 30, 1
   6, 20, 17, 1, 21, 13, 22, 24, 19, 30, 12, 24, 29, 18, 26, 23, 22, 31

2. 17, 50, 46, 49, 18, 27, 20, 29, 45, 42, 30, 53, 32, 54, 33, 39, 36, 64
   40, 62, 41, 43, 26, 22, 35, 44, 61, 52, 34, 59, 36, 57, 38, 58, 23, 47

3. 73, 48, 64, 49, 55, 57, 68, 50, 49, 78, 59, 82, 50, 81, 54, 65, 84, 53
   57, 70, 59, 83, 71, 54, 75, 78, 53, 80, 77, 65, 69, 52, 58, 58, 61, 56

12-4 Study Guide and Intervention

Comparing Sets of Data

Transformation of Data When an operation is performed on every value of a data set, the statistics for the new set of data can be found using the statistics from the original set of data.

Transformations Using Addition

If a real number k is added to every value in a set of data, then:

- the mean, median, and mode of the new data set can be found by adding k to the mean, median, and mode of the original data set, and
- the range and standard deviation will not change.

Transformations Using Multiplication

If every value in a set of data is multiplied by a constant k, $k > 0$, then the mean, median, mode, range, and standard deviation of the new data set can be found by multiplying each original statistic by k.

Example 1 Find the mean, median, mode, range, and standard deviation of the data set obtained after adding 9 to each value.

12, 10, 15, 17, 15, 9, 10, 15, 12, 14

The mean, median, mode, range, and standard deviation of the original data set are 12.9, 13, 15, 8, and about 2.5, respectively. Add 9 to the mean, median, and mode. The range and standard deviation are unchanged.

Mean 21.9   Median 24   Mode 22

Range 8   Standard Deviation 2.5

Example 2 Find the mean, median, mode, range, and standard deviation of the data set obtained after multiplying each value by 1.5.

4, 3, 7, 6, 2, 6, 8, 5, 4, 6, 7, 2

The mean, median, mode, range, and standard deviation of the original data set are 5, 5.5, 6, 6, and about 1.9, respectively. Multiply the mean, median, mode, range, and standard deviation by 1.5.

Mean 7.5   Median 8.3   Mode 9

Range 9   Standard Deviation 2.9

Exercises

Find the mean, median, mode, range, and standard deviation of each data set that is obtained after adding the given constant to each value.

1. 33, 38, 29, 35, 27, 34, 36, 28, 41, 26; + 11   **2.** 8, 9, 3, 6, 12, 9, 3, 16, 9, 11; + (−3)

Find the mean, median, mode, range, and standard deviation of each data set that is obtained after multiplying each value by the given constant.

3. 2, 1, 8, 6, 3, 1, 7, 5, 7, 2, 4, 2; × 4   **4.** 22, 26, 30, 27, 25, 23, 31, 20; × 0.5

12-4 Study Guide and Intervention (continued)

Comparing Sets of Data

Comparing Distributions When comparing two sets of data, use

- the means and standard deviations if the distributions are both symmetric, or

- the five-number summaries if the distributions are both skewed, or if one distribution is symmetric and the other is skewed.

Example **ATTENDANCE** The attendance for each of the PTSA meetings at the two elementary schools in Gahanna Heights School District is shown.

Rocky Run Elementary	Trail Woods Elementary
20, 56, 25, 45, 41, 27, 28, 51, 30, 34, 23, 37	76, 63, 57, 69, 50, 54, 40, 67, 36, 65, 74, 28

a. Use a graphing calculator to construct a box-and-whisker plot for each set of data. Then describe the shape of each distribution.

Enter Rocky Run's attendance as L1 and Trail Woods' as L2. Graph both box-and-whisker plots on the same screen.

For Rocky Run, the distribution is positively skewed. For Trail Woods, the distribution is negatively skewed.

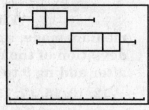

[15, 80] scl: 5 by [0, 5] scl: 1

b. Compare the data sets using either the means and standard deviations or the five-number summaries. Justify your choice.

Both distributions are skewed, so use the five-number summaries to compare the data.

The maximum for Rocky Run is 56, while the median for Trail Woods is 60. This means that at half of the meetings at Trail Woods, the attendance was greater than at any of the meetings at Rocky Run. We can conclude that overall, the attendance at Trail Woods' meetings was greater than the attendance at Rocky Run's meetings.

Exercise

SWIMMING Gracie's times in the 50-yard freestyle over two years are shown.

Sophomore Year (seconds)	Junior Year (seconds)
24.5, 25.7, 24.9, 25.3, 25.8, 25.9, 26.1, 26.3, 26.4, 26.6, 26.9, 27.5	24.2, 24.4, 24.5, 24.6, 24.6, 24.9, 25.1, 24.1, 24.1, 23.9, 23.7, 23.5

a. Use a graphing calculator to construct a box-and-whisker plot for each set of data. Then describe the shape of each distribution.

b. Compare the data sets using either the means and standard deviations or the five-number summaries. Justify your choice.

12-5 Study Guide and Intervention

Simulations

Experimental Probability A **theoretical probability** is the ratio of the number of feasible outcomes to the total number of outcomes. For example, the theoretical probability of rolling a 4 with a number cube is $\frac{1}{6}$, or $P(4) = \frac{1}{6}$. **Experimental probability** is the ratio of the number of times an outcome occurs in an experiment to the total number of events or trials, known as the **relative frequency**.

Experimental Probability	$P = \dfrac{\text{frequency of an outcome}}{\text{total number of trials}}$

Example 1 Matt recorded that it rained 8 times in November and snowed 3 times. The other days, it was sunny. There are 30 days in November. Find the experimental probability that a day in November is sunny.

$$\text{Experimental Probability} = \frac{\text{frequency of outcome}}{\text{total number of trials}}$$
$$= \frac{(30 - 8 - 3)}{30}$$
$$= \frac{19}{30} = 63.3\%$$

The experimental probability that it was sunny on a day in November is 63.3%.

Example 2 A football team noticed that 9 of the last 20 coin tosses to choose which team would receive the ball first resulted in tails. What is the experimental probability of the coin landing on tails? What is the theoretical probability?

$$\text{Experimental Probability} = \frac{\text{frequency of outcome}}{\text{total number of trials}}$$
$$= \frac{\text{number of tails}}{\text{total number of tossess}}$$
$$= \frac{9}{20}$$

In this case, the experimental probability that a coin toss will be tails is $\frac{9}{20}$ or 45%. If the coin is fair, the theoretical probability is $\frac{1}{2}$ or 50%.

Exercise

DIE ROLL A math class decided to test whether a die is fair, that is, whether the experimental probability equals the theoretical probability. The results for 100 rolls are shown at the right.

1: 6	2: 15
3: 4	4: 13
5: 15	6: 47

a. What is the theoretical probability of rolling a 6?

b. What is the experimental probability of rolling a 6?

c. Is the die fair? Explain your reasoning.

12-5 Study Guide and Intervention *(continued)*

Simulations

Simulations A **simulation** allows you to use objects to act out an event that would be difficult or impractical to perform.

Use the following steps to design a simulation.

Step 1	Determine each possible outcome and its theoretical probability.
Step 2	Describe an appropriate probability model for the situation that accurately represents the theoretical probability of each outcome.
Step 3	Define what a trial is for the situation, and state the number of trials to be conducted.

Example Pete got a hit 84 of his last 252 at bats. Design a simulation that can be used to estimate the probability that Pete will get a hit at his next at bat.

There are two possible outcomes: a hit and not a hit.

$P(\text{hit}) = \dfrac{84}{252}$ or $\dfrac{2}{6}$

We can use a spinner like the one at the right with 6 equally likely outcomes. Let sections 1 and 4 represent Pete getting a hit, and sections 2, 3, 5, and 6 represent Pete not getting a hit. The simulation can consist of any number of trials. We will use 40.

Exercises

1. **GUESSING** Design a simulation that can be used to estimate the probability of guessing an answer correctly on a true or false test.

2. **PROMOTIONS** Main Street Market randomly gives each shopper a two-liter bottle of one of four types of cola during a sale. Design a simulation that can be used to estimate the probability of each type of cola being given away.

3. **PICNICS** At a picnic, there are two peanut butter sandwiches, two chicken sandwiches, a tuna sandwich, and a turkey sandwich in a cooler. Describe a simulation that can be used to estimate the probability of picking a certain sandwich from the cooler.

12-6 Study Guide and Intervention

Permutations and Combinations

Permutations An arrangement or listing in which order or placement is important is called a **permutation.** For example, the arrangement AB of choices A and B is different from the arrangement BA of these same two choices.

Permutations	The number of permutations of n objects taken r at a time is $$P(n, r) = \frac{n!}{(n - r)!}.$$

Example 1   Find $P(6, 2)$.

$P(n, r) = \dfrac{n!}{(n - r)!}$ Permutation Formula

$P(6, 2) = \dfrac{6!}{(6 - 2)!}$ $n = 6, r = 2$

$= \dfrac{6!}{4!}$ Simplify.

$= \dfrac{6 \cdot 5 \cdot 4 \cdot 3 \cdot 2 \cdot 1}{4 \cdot 3 \cdot 2 \cdot 1}$ Definition of factorial

$= 6 \cdot 5$ or 30   Simplify.

There are 30 permutations of 6 objects taken 2 at a time.

Example 2   **PASSWORDS A specific program requires the user to enter a 5-digit password. The digits cannot repeat and can be any five of the digits 1, 2, 3, 4, 7, 8, and 9.**

a. How many different passwords are possible?

$P(n, r) = \dfrac{n!}{(n - r)!}$

$P(7, 5) = \dfrac{7!}{(7 - 5)!}$

$= \dfrac{7 \cdot 6 \cdot 5 \cdot 4 \cdot 3 \cdot 2 \cdot 1}{2 \cdot 1}$

$= 7 \cdot 6 \cdot 5 \cdot 4 \cdot 3$ or 2520

There are 2520 ways to create a password.

b. What is the probability that the first two digits are odd numbers with the other digits any of the remaining numbers?

$P(\text{first two digits odd}) = \dfrac{\text{number of favorable outcomes}}{\text{number of possible outcomes}}$

favorable outcomes: There are 4 choices for the first 2 digits and 5 choices for the remaining 3 digits. $P(4, 2) \cdot P(5, 3)$

possible outcomes: There are 7 choices for the 5 digits. $P(7, 5)$

The probability is $\dfrac{P(4, 2) \cdot P(5, 3)}{P(7, 5)} = \dfrac{720}{2520}$ or about 28.6%.

Exercises

Evaluate each expression.

1. $P(7, 4)$ **2.** $P(12, 7)$ **3.** $[P(9, 9)]$

4. CLUBS A club with ten members wants to choose a president, vice-president, secretary, and treasurer. Six of the members are women, and four are men.

  a. How many different sets of officers are possible?

  b. What is the probability that all officers will be women.

12-6 Study Guide and Intervention (continued)

Permutations and Combinations

Combinations An arrangement or listing in which order is not important is called a combination. For example, AB and BA are the same combination of A and B.

Combinations	The number of combinations of n objects taken r at a time is $$C(n, r) = \frac{n!}{(n-r)!\, r!}.$$

Example A club with ten members wants to choose a committee of four members. Six of the members are women, and four are men.

a. How many different committees are possible?

$C(n, r) = \dfrac{n!}{(n-r)!\, r!}$ Combination Formula

$C(10, 4) = \dfrac{10!}{(10-4)!\, 4!}$ $n = 10, r = 4$

$ = \dfrac{10 \cdot 9 \cdot 8 \cdot 7}{4!}$ Divide by the GCF 6!.

$ = 210$ Simplify.

There are 210 ways to choose a committee of four when order is not important.

b. If the committee is chosen randomly, what is the probability that two members of the committee are men?

Probability (2 men and 2 women) $= \dfrac{\text{number of favorable outcomes}}{\text{number of possible outcomes}}$

favorable outcomes: There are 4 choices for the 2 men $C(4, 2) \cdot C(6, 2)$
and 6 choices for the 2 remaining spots.

The probability is $\dfrac{C(4, 2) \cdot C(6, 2)}{C(10, 4)} = \dfrac{90}{210}$ or about 42.9%.

Exercises

Evaluate each expression.

1. $C(7, 3)$ **2.** $C(12, 8)$ **3.** $C(9, 9)$

4. COMMITTEES In how many ways can a club with 9 members choose a two-member sub-committee?

5. BOOK CLUBS A book club offers its members a book each month for a year from a selection of 24 books. Ten of the books are biographies and 14 of the books are fiction.

   a. How many ways could the members select 12 books?

   b. What is the probability that 5 biographies and 7 fiction books will be chosen?

12-7 Study Guide and Intervention

Probability of Compound Events

Independent and Dependent Events **Compound events** are made up of two or more simple events. The events can be **independent events** or they can be **dependent events**.

Probability of Independent Events	Outcome of first event does not affect outcome of second.	$P(A \text{ and } B) = P(A) \cdot P(B)$	Example: rolling a 6 on a die and then rolling a 5
Probability of Dependent Events	Outcome of first event does affect outcome of second.	$P(A \text{ and } B) = P(A) \cdot P(B \text{ following } A)$	Example: without replacing the first card, choosing an ace and then a king from a deck of cards

Example 1 Find the probability that you will roll a six and then a five when you roll a die twice.

By the definition of independent events,
$P(A \text{ and } B) = P(A) \cdot P(B)$

First roll: $P(6) = \dfrac{1}{6}$

Second roll: $P(5) = \dfrac{1}{6}$

$P(6 \text{ and } 5) = P(6) \cdot P(5)$
$= \dfrac{1}{6} \cdot \dfrac{1}{6}$
$= \dfrac{1}{36}$

The probability that you will roll a six and then roll a five is $\dfrac{1}{36}$.

Example 2 A bag contains 3 red marbles, 2 green marbles, and 4 blue marbles. Two marbles are drawn randomly from the bag and not replaced. Find the probability that both marbles are blue.

By the definition of dependent events,
$P(A \text{ and } B) = P(A) \cdot P(B \text{ following } A)$

First marble: $P(\text{blue}) = \dfrac{4}{9}$

Second marble: $P(\text{blue}) = \dfrac{3}{8}$

$P(\text{blue, blue}) = \dfrac{4}{9} \cdot \dfrac{3}{8}$
$= \dfrac{12}{72}$
$= \dfrac{1}{6}$

The probability of drawing two blue marbles is $\dfrac{1}{6}$.

Exercises

A bag contains 3 red, 4 blue, and 6 yellow marbles. One marble is selected at a time, and once a marble is selected, it is not replaced. Find each probability.

1. $P(2 \text{ yellow})$ **2.** $P(\text{red, yellow})$ **3.** $P(\text{blue, red, yellow})$

4. George has two red socks and two white socks in a drawer. What is the probability of picking a red sock and a white sock in that order if the first sock is not replaced?

5. Phyllis drops a penny in a pond, and then she drops a nickel in the pond. What is the probability that both coins land with tails showing?

6. A die is rolled and a penny is dropped. Find the probability of rolling a two and showing a tail.

12-7 Study Guide and Intervention (continued)

Probability of Compound Events

Mutually Exclusive and Inclusive Events Events that cannot occur at the same time are called **mutually exclusive**. If two events are not mutually exclusive, they are called **inclusive**.

Probability of Mutually Exclusive Events	$P(A \text{ or } B) = P(A) + P(B)$	$P(\text{rolling a 2 or a 3 on a die}) = P(2) + P(3) = \frac{1}{3}$
Probability of Inclusive Events	$P(A \text{ or } B) =$ $P(A) + P(B) - P(A \text{ and } B)$	$P(\text{king or heart}) = P(K) + P(H) - P(K \text{ and } H) = \frac{9}{26}$

Example A card is drawn from a standard deck of playing cards.

a. Find the probability of drawing a king or a queen.

Drawing a king or a queen are mutually exclusive events. Use the formula for the probability of mutually exclusive events, $P(A \text{ or } B) = P(A) + P(B)$.

$P(A) = P(\text{king}) = \frac{4}{52} \text{ or } \frac{1}{13}$ $P(B) = P(\text{queen}) = \frac{4}{52} \text{ or } \frac{1}{13}$

$P(\text{king or queen}) = \frac{1}{13} + \frac{1}{13} \text{ or } \frac{2}{13}$

The probability of drawing a king or a queen is $\frac{2}{13}$.

b. Find the probability of drawing a king or a spade.

Since it is possible to draw a card that is a king and a spade, these are inclusive events. Use the formula $P(A \text{ or } B) = P(A) + P(B) - P(A \text{ and } B)$.

$P(A) = P(\text{king}) = \frac{4}{52}$ $P(B) = P(\text{spade}) = \frac{13}{52}$ $P(A \text{ and } B) = P(\text{king of spades}) = \frac{1}{52}$

$P(\text{king or spade}) = \frac{4}{52} + \frac{13}{52} - \frac{1}{52} = \frac{16}{52} \text{ or } \frac{4}{13}$

The probability of drawing a king or a spade is $\frac{4}{13}$.

Exercises

A bag contains 2 red, 5 blue, and 7 yellow marbles. Find each probability.

1. $P(\text{yellow or red})$ **2.** $P(\text{red or not yellow})$ **3.** $P(\text{blue or red or yellow})$

A card is drawn from a standard deck of playing cards. Find each probability.

4. $P(\text{jack or red})$ **5.** $P(\text{red or black})$ **6.** $P(\text{jack or clubs})$

7. $P(\text{queen or less than 3})$ **8.** $P(5 \text{ or } 6)$ **9.** $P(\text{diamond or spade})$

10. In a math class, 12 out of 15 girls are 14 years old and 14 out of 17 boys are 14 years old. What is the probability of selecting a girl or a 14-year old from this class?

12-8 Study Guide and Intervention
Probability Distributions

Random Variables and Probability A **random variable** X is a variable whose value is the numerical outcome of a random event.

Number of Siblings	Number of Students
0	1
1	15
2	8
3	2
4	1

Example A teacher asked her students how many siblings they have. The results are shown in the table at the right.

a. Find the probability that a randomly selected student has 2 siblings.

The random variable X can equal 0, 1, 2, 3, or 4. In the table, the value $X = 2$ is paired with 8 outcomes, and there are 27 students surveyed.

$$P(X = 2) = \frac{2 \text{ siblings}}{27 \text{ students surveyed}}$$
$$= \frac{8}{27}$$

The probability that a randomly selected student has 2 siblings is $\frac{8}{27}$, or 29.6%.

b. Find the probability that a randomly selected student has at least three siblings.

$$P(X \geq 3) = \frac{2 + 1}{27}$$

The probability that a randomly selected student has at least 3 siblings is $\frac{1}{9}$, or 11.1%.

Exercises

For Exercises 1–3, use the grade distribution shown at the right. A grade of A = 5, B = 4, C = 3, D = 2, and F = 1.

X = Grade	5	4	3	2	1
Number of Students	6	9	5	4	1

1. Find the probability that a randomly selected student in this class received a grade of C.

2. Find the probability that a randomly selected student in this class received a grade lower than a C.

3. What is the probability that a randomly selected student in this class passes the course, that is, gets at least a D?

4. The table shows the results of tossing 3 coins 50 times. What is the probability of getting 2 or 3 heads?

X = Number of Heads	0	1	2	3
Number of Times	6	20	19	5

12-8 Study Guide and Intervention (continued)

Probability Distributions

Probability Distributions The probabilities for every possible value of the random variable X make up its **probability distribution**.

Properties of a Probability Distribution	1. The probability of each value of X is greater than or equal to 0.
	2. The probabilities for all values of X add up to 1.

The probability distribution for a random variable can be given in a table or a **probability graph** and used to obtain other information. The **expected value $E(X)$** of a discrete random variable of a probability distribution is the weighted average of the variable.

Example Use the data in the table below to determine a probability distribution and to make a probability graph.

Number of Siblings	Number of Students
0	1
1	15
2	8
3	2
4	1

X = Number of Siblings	$P(X)$
0	0.037
1	0.556
2	0.296
3	0.074
4	0.037

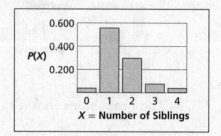

a. Show that the distribution is valid.

For each value of X, $0 \le P(X) \le 1$. The sum of the probabilities is 1.

b. What is the probability that a student has fewer than 2 siblings?

$P(X < 2) = 0.037 + 0.556$ or 0.593

c. What is the expected number of siblings for a randomly selected student?

$E(X) = [X_1 \cdot P(X_1)] + [X_2 \cdot P(X_2)] + \dots + [X_n \cdot P(X_n)]$

$= 0(0.037) + 1(0.556) + 2(0.296) + 3(0.074) + 4(0.037)$

$= 0 + 0.556 + 0.592 + 0.222 + 0.148$ or 1.518

The expected number is 1.518, so a student can be expected to have 1 or 2 siblings.

Exercises

The table at the right shows the probability distribution for the ratings of an amusement park ride.

X = Rating	$P(X)$
1 = Very Poor	0.096
2 = Poor	0.143
3 = Average	0.342
4 = Good	0.318
5 = Excellent	0.101

1. Show that the distribution is valid.

2. If a rating is chosen at random, what is the probability that it is less than 4?

3. Make a probability graph of the data.

4. What is the expected value of a randomly selected rating?

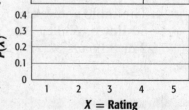